The Search for Water

VIRGINIA MENDOZA studied Journalism and Social and Cultural Anthropology. She works as a freelance journalist, writing for various media in Spain and Latin America, and she has published books on rootedness and uprooting in which she fuses narrative journalism and rural anthropology. *The Search for Water*, a hybrid of essay, historical anthropology and memoir, is being translated into several languages.

THOMAS BUNSTEAD was born in London in 1982 and lives in Pembrokeshire, west Wales. He has translated some of the leading Spanish-language writers at work today, including Maria Gainza, Agustín Fernández Mallo and Enrique Vila-Matas, and won various awards, including an O. Henry Prize in 2023 and the 2024 Cercador Prize for Literature in Translation.

The Search for Water

Virginia Mendoza

TRANSLATED FROM THE SPANISH
BY THOMAS BUNSTEAD

PUSHKIN PRESS

Pushkin Press
Somerset House, Strand
London WC2R 1LA

Original text © 2024, Virginia Mendoza Benavente
English translation © 2026, Thomas Bunstead

The Search for Water was first published as *La sed* by Penguin
Random House Grupo Editorial in Barcelona, 2024

First published by Pushkin Press in 2026

The right of Virginia Mendoza to be identified as the author of this Work has been
asserted by them in accordance with the Copyright, Designs & Patents Act 1988

ISBN 13: 978-1-80533-293-0

Epigraph to Chapter 4 quoted from *Origins* by Lewis Dartnell,
published by Bodley Head. Copyright © Lewis Dartnell, 2019.
Reprinted by permission of The Random House Group Limited.

Extracts from *The Shadow of the Sun* by Ryszard Kapuscinski
published by Penguin. Copyright © Ryszard Kapuscinski 1998.
Reprinted by permission of Penguin Books Limited.

All rights reserved. No part of this publication may be reproduced,
stored in a retrieval system or transmitted in any form or by any
means, electronic, mechanical, photocopying, recording or otherwise,
or for the purpose of training artificial intelligence technologies or
systems without prior permission in writing from Pushkin Press

A CIP catalogue record for this title is available from the British Library

The authorised representative in the EEA is eucomply OÜ,
Pärnu mnt. 139b-14, 11317, Tallinn, Estonia,
hello@eucompliancepartner.com, +33757690241

Designed and typeset by Tetragon, London
Printed and bound in the United Kingdom by Clays Ltd, Elcograf S.p.A.

Pushkin Press is committed to a sustainable future for our
business, our readers and our planet. This book is made from
paper from forests that support responsible forestry.

www.pushkinpress.com

1 3 5 7 9 8 6 4 2

To Dani,
who got me writing again.

To my grandmother Francisca,
who appears here in both the present and past tense
and who left the final question unanswered.

To my parents and brother,
alongside whom I experienced thirst.

In memory of Fati and Marie, mother and daughter,
who died of thirst in the middle of the Libyan
desert while I was finishing this book.

CONTENTS

Prologue   11

    I   THE JOURNEY OF THIRST

1   Heavenly pork   31
2   *Homo sitibundus*: the great journey   63
3   Water lessons   81
4   Waiting for rain   93
5   Under dry ground   125

    II   MANAGING THE RAINS

6   Heavenly horns   161
7   God came to earth   191
8   Rainmaker   211
9   Drought and disorder   235
10   Feet on the ground, eyes on the sky   261

Epilogue: Exodus of the thirsty   285
Acknowledgements   309
Select Bibliography   313

> Of what ancient desert are you a memory,
> you who thirst, you who consume yourself in water,
> lifting up your dead body into space
> as if your water were heaven-sent?
>
> ALFONSINA STORNI

It cannot be, señor, but that this grass is a proof that there must be hard by some spring or brook to give it moisture, so it would be well to move a little farther on, that we may find some place where we may quench this terrible thirst that plagues us, which beyond a doubt is more distressing than hunger.

MIGUEL DE CERVANTES,
THE HISTORY OF DON QUIXOTE

> Of course God exists.
> It is a woman
> and her name is Rain.
>
> GUSTAVO DUCH

PROLOGUE

> Not much time had passed when I realised I was thirsty and had no water. I wanted to wait before I went looking for some, but then I remembered that such things as thirst, death and love are inescapable. Sooner or later I would have to go and get some.
>
> NÚRIA BENDICHO GIRÓ,
DEAD LANDS

I neither can nor want to forget the place in La Mancha where I first came to know the human search for water. An old bath, surrounded by pots and pans, stood out in my maternal grandparents' farmyard awaiting rain. The nearby Villanueva River was running dry and had ceased to irrigate the orchards and vegetable patches of Villanueva de la Fuente, Ciudad Real. Crops were failing and one local woman had been forced to sell her cows. Reserves were also down; Aquifer 24, or the Campo de Montiel Aquifer, from which the river flowed, had virtually dried up. Although they were told it was because of the rain, or the lack of it, local growers had begun to suspect that something else was at play. At the height of a drought, while their crops perished, the corn on a nearby duke's estate was growing splendidly across an area of almost a thousand hectares, with the help of a modern irrigation system. In August 1987, the people of Villanueva de la Fuente and nearby Albaladejo, Villahermosa and Montiel organized a march. They showed up at the estate with upturned drinking jugs and placards declaiming: "We're thirsty!" and "We want our water back!" But nothing changed.

August 15th that year was a Saturday, and some Villanuevans, certain by now that the drought had little to do with the absence of rain, went and knocked down four of the poles that carried electricity into the corn-growing duke's estate. On the Sunday morning, when they saw workers out there doing repairs, they went and toppled the same four again, and nineteen more besides. Who was responsible? "It was all of us, sir," they said. The village numbered about 3,500 inhabitants throughout the year, and many more, twice as many, in the middle of August. They performed their own *Fuenteovejuna*, Lope de Vega's play of uprising against unjust rule, though in this version no blood was spilt.

When the journalist Luis Otero went to interview the local people, he was told: "There's no ringleader here, if that's what you want to know. The village is all of us, together. And if we happen to hear they've come to put up new power lines, we'll be on them like an avalanche, and we'll put a stop to it. But only with our hands, no weapons. We don't want violence, we only want back what's ours. That is, the water." Otero had gone there asking after the woman who had sold her cows. Her name was Julia, but the other villagers had started referring to her as Agustina of Aragón. She was an old woman who composed protest ballads, and who emerged as both a figurehead and chronicler of her community's revolt.

A quote from another Villanuevan in *El País* sums up events in the village:

"The water's always been ours, all God's life it's been ours, until that man diverted it to water his corn."

They blamed the duke's son, who'd had a series of 150m-deep wells dug and connected to a sophisticated irrigation system, siphoning off water that belonged to everyone. But they had also long harboured suspicions about the farmer on the neighbouring estate:

"Yes, we said, there was drought, but it was the estates wrecking the springs and the Ruidera lagoons." So said Juan Ángel Amador when I went to speak to him—the man who had just been appointed mayor when the 1987 conflict blew up.

The riot police soon arrived, reportedly some 200 of them. But the mutinous locals got their way, and after the mayor intervened to prevent the poles from being repaired, they ended up going out and giving the riot police a mocking round of applause. And the river began to flow once more. Justice was done and the people were proved right; two years later, the aquifer was declared overexploited.

Riot police had made an appearance in another village that summer. While the people of Villanueva had stopped electricity poles being repaired, those of Riaño, in northerly León, had climbed onto the roofs of their houses and refused to get down. This was their desperate, ultimately vain attempt to resist eviction; Riaño, along with eight other villages in the area, had been designated for flooding as part of a new reservoir intended for irrigation and hydroelectricity.

Photographs in that summer's newspapers show that those in search of water and those drowned by it are often two sides of the same coin. While a number of children went to protest in

Villanueva's dried-up riverbed, one climbed onto his roof to protest against the flooding of his village. Those newspaper images memorialized them all.

People continued to go thirsty, as another drought soon followed. In Spain and other Mediterranean countries, droughts are cyclical; every decade or so, one lasting three or four years takes place. In the summer of 1992, when Spain was divided between those who slept during siesta and those who followed Miguel Induráin's attempt to win the Tour de France for a second time, in Terrinches, my native village, water was once again just about the only thing on our minds. The water that wouldn't come; water that, if it continued to stay away, would mean we would have to leave. The elderly were particularly desperate, and it was then that I came to value water in the way that only lost things are valued. For some time it became a thing of mystery, arriving as it did on the back of tanker trucks or by dint of my grandfather Norberto's labours. The underground tanks that were dug during that time are still there, in case of further droughts.

Because it was normal for water to be absent, moments in which it did feature have stayed with me particularly strongly. My grandfather at the back of the cave hunting for drips to redirect to a tank that would in turn send the water on to the orchards and vegetable patches. My grandfather going from the orchards to the farmyard for a stripwash out of a bucket. Shared family baths, every single drop maximized and recycled. The only thing we didn't wring out was the air. Everything served as a way to retain the almost nonexistent precipitation, which would then be stored like treasure, even when it was good for almost nothing. Perhaps this explains the very clear image I still have of some tadpoles that were born and proliferated inside a petrol canister. That drought, which lasted until 1995, left Spain's reservoirs at fifteen per cent, and dried out the well my village had relied on for centuries. While my neighbours went to another village to pray to all the saints for the rain, elsewhere people were discussing getting tugboats to bring an iceberg up the Guadalquivir River to supplement the flow. It was either that or transplant the entire population of Seville. The iceberg idea was not a new one; it had been discussed in Benidorm during another drought almost two decades before.

Antonio Muñoz Molina's novel *El viento de la luna* (Moon Wind), whose main character is a boy captivated by the first moon landings, is set in a village in Jaén very near the one I grew up in, and one just as parched. Pedro, the protagonist's uncle, has the mad idea of installing a shower in the farmyard. "But the only way to have a wash here," he says, "is to get a bucket of icy water from the well and pour it through a cracked washbowl. In these harsh lands, running water is a dream as distant as that of timely, abundant rain." We had our own lunatic in Terrinches who also claimed to have a shower in his farmyard at a time when nobody had any running water. Muñoz Molina's cracked washbowl and other devices rigged up during droughts are so familiar to me, it's almost as though I grew up in that same house. Although his story takes place thirty years earlier and in another place, it is also my own; a farmyard full of pots and pans rather than chickens. I even have photos of the first bath I ever took on my own, not because it was a big moment in my development but because it was a luxury that had to be immortalized like anything else suspected to be a one-off.

My grandfather was the person who looked after the water in our village. As well as being the street sweeper, tree planter, and the crier—announcing deaths and cutting the rosary that was tied around a dead person's hands until just before they were lowered into the ground—he was charged with caring for those who were alive and thirsty, and held the key to the sluice inside the harvesting tank. I often went with him for this job. After siesta, I would watch as he descended the metal steps into the underworld to cut off the village supply with a turn of the key. This was a novelty, in a way. Running water came late to Terrinches. There, only the figures of Don Quixote and the Virgin of Luciana were comparable in veneration to the drinking jug with which they still formed a trinity. Occupying a nook that was virtually an altar, the drinking jug was a thing of awe. So as not to spill a drop of the water brought from the spring, and to stop the flies from pilfering it, my grandmother would put the jug on a plate of its own and fasten a made-to-measure crochet lid to the top with a ribbon. Our history is determined by our relationship with water. In our connection with it, there is always the lurking fear that at some point it will abandon us again.

I've been told that in that summer of 1992, days as dry as salted tuna, my grandfather sometimes turned the key to let the village have water for just thirty minutes out of the twenty-four-hour cycle. Then everyone would be dashing for a shower, to wash the dishes, have a drink. Sometimes there wasn't long enough even for one of the rapid cycles on the washing machine, and it was my mother and aunts who went to open and close the sluice while their father went around the village to let everyone know. I don't know if it happened in all that rushing around, but my grandfather lost half of one of his little fingers in the door to the tank, and every time he cut the bread with his knife or lifted the wineskin or water jug, I'd be confronted with the sight of the nub pointing either up at the ceiling or right at me. We used to make fun of my grandmother because she couldn't bring herself to use the washing machine and would leave it mothballed, going on washing clothes by hand with her own oil-and-soda soap. I now understand that people's reverence for washing machines is not only based on the fear of their blowing up or breaking if you use them.

That year, home videos were in vogue, and in another part of Spain—the Atlantic northwest, with its far wetter climate—the village of Aceredo in Galicia was flooded to create a reservoir. Among the onlookers was the astonished resident Paco Villalonga, who recorded everything on his camcorder, there being nothing else he could do. Others had barricaded themselves inside the town hall and written banners with messages like, "We are on hunger strike to show the dignity that the Spanish government lacks!" and "Death and destruction to 200 working families. Human rights violations. Listen to us!" But the government didn't listen. I knew nothing of any of that when I was five years old, but some time afterwards I spoke to Paco, who told me that every time the reservoir now runs low, exposing what remains of the otherwise completely submerged village, he would go to the ruins of his old house and eat a sandwich by a font that continued to send out water. It was a strange sight; being under the immense body of water most of the year had done nothing to alter the natural flow of the springs there.

Now that I have come to ask questions about that time, I have learned that in the end a farmer (one of those accused in Villanueva) gave us access to one of his wells. A grant from the Board of Public Works paid for the channelling and supply works to carry the water over a stretch of some twenty kilometres, and since 1995 Terrinches has in large part been drinking from that source. People tell the story with gratitude. But the concession agreement, signed at the end of August that year, concludes with a statement that the authorization may be cancelled "whenever deemed appropriate for any reason, which in no case will have to be justified, with two months' notice to the beneficiary municipality being sufficient, without the latter being able to oppose it or claim any compensation for any reason whatsoever". So the question of a village going without water depends almost exclusively on the will of one man, or rather on something that does not in fact exist: the will of a corporation.

Ours is the thirst of all of Spain's dry places—three quarters of the entire peninsula, in other words, which is steppe- and desert-like in places. And our search is the very same as that undertaken by our most distant ancestors. In this part of Castilla-La Mancha we do not even get half of the 400 litres of rain per m^2 that is the median in the region overall. Leaving because there is no water; leaving because the water has come: this is a country of the thirsty and those drowned because of thirst. And this is the story we forget when we turn the tap on, one that has been encoded in our genes. But it is also a story from before, from far away, concerning all human inhabitants of planet Earth. Our family, our genus and our species came into being with the transfer of peoples from East Africa across arid mountains. If the earliest human fossils have been found in the middle and lower reaches of an African river, the Awash, the first ever civilizations also emerged alongside rivers in the midst of droughts. Throughout history the search for water has driven great anatomical and metabolic adaptations, as well as social innovations and revolutions, and indeed the collapse of civilizations. In the coming pages we will see how almost everything that defines our species came about in the face of climate change, with the alternation of dry and wet periods. There is no reason for us to be surprised by the latest climate crisis; we are the children of climate crises. Perhaps any such surprise betrays a measure of guilt.

This story takes place in the Cenozoic, the era that gave rise to both our direct ancestors and the staples of our modern diet. It begins with Lucy in the Neogene, but the principal time period is the Quaternary, which includes the present day. This period encompasses two epochs, the Pleistocene and the Holocene—the latter also a word for the period we are currently in—which were divided, indeed, by changes in climate. In all this time, tens of millions of years, there have been several cold-dry and warm-wet cycles. For climatic periods are like Russian dolls; although we are now in a warming phase, more broadly the world has been cooling and drying for about fifty million years, a paradox that unwittingly fuels climate-denialism. Instability has led to successive disruptions within the overall trend. About 2.6 million years ago, the world entered a constant cycle of ice ages and interglacial periods; it was then that humanity also emerged. We have been in an interglacial epoch for the last 11,700 years, though this has also included colder phases. In short, strange as it may seem, the Earth is warming and cooling at the same time. This is largely because, especially over the last 300 years, our presence has disrupted the planet's natural post-Neolithic trends.

Although some of the most relevant changes in climate under consideration here have been linked to extra-terrestrial causes like comets exploding or sunspot shrinkage, we will see that they came about primarily due to astronomical movements related to the Earth's position with respect to the Sun, the shape of its orbit and the tilt of its axis of rotation. In addition, during this time there have been changes in climate for geological reasons, such as the movement of tectonic plates, earthquakes, volcanic eruptions and alterations in ocean currents. These causes have often converged, and our climate system depends on several factors, such as the atmosphere which, as well as allowing us to breathe, maintains an average temperature of 15°C through "greenhouse" gases which balance the energy received and emitted by the Earth but which we have artificially increased, contributing to global warming. Ocean currents, too, are part of this balance, through their interaction with the atmosphere. And finally, there is solar radiation. To all this we must add a new trigger: we humans and our actions.

The climate has pushed us to the brink of extinction before: we are the descendants of the handful of hominids (some 1,300) who survived extreme cold and aridity less than 200,000 years ago. And although only we *Homo sapiens* came through the last ice age, we didn't come through unscathed. Yet it wasn't until 1988 that climate change became a matter of concern beyond purely scientific circles. Desperation at the soaring temperatures inside the US Senate building finally led to global warming becoming a public concern. In places like Spain, none the less, talk of the weather continued to be frowned upon, and it is still considered an innocuous conversational gambit for surviving rides in lifts with strangers. But the climate, seemingly so insignificant, is one of the very reasons why some of our ancestors came to the place where we were born, and long before that, why theirs in turn had to leave Africa.

We cannot go from ignoring the climate to denying its variations, because this is like renouncing LUCA (the Last Universal Common Ancestor) just because we don't like the idea of descending from a bacterium. Changes in climate have always been with us, spurring everything from evolution to migration, to cultural innovation and the mixing of genes. They are part of us, as we are part of them. Cognitive developments laid the first stone in the freedom we enjoy today, but freedom implies responsibility. Culture promised us, with nature's blessing, an independence that appeared absolute. But it wasn't so. There is no such thing as "crazy" weather, and avoiding our responsibility can only make freedom less attainable, and us in turn more vulnerable. Nor is there any point in succumbing to pessimism; a pessimist, believing change impossible, has decided not even to try. We can only let ourselves be driven by optimism, the rational rather than the blind variety; not the idea of a divine plan but the will to fix what we have broken, in the knowledge that there are still some pieces that can be repaired. No action without hope. We must do it in the only way that has led to good outcomes in our history: collectively. To do so, we need to recover our awareness of ourselves as a single species, without losing sight of the fact that we are also one with nature and that not everyone is in a position to leave the same footprint and, therefore, to reduce it.

According to a report by the Spanish Hydrographic Studies Centre, all the indications are that the wet parts of Spain—which have some of the wettest spots in Europe—will remain wet even if rainfall decreases, and that the dry parts of Spain—which include some of the continent's very driest spots—will become even drier. The European Environment Agency forecasts that, in all of Europe, the Iberian peninsula will be the area most affected by desertification in the coming years. Unregulated irrigation, overexploitation of aquifers, soil degradation and land abandonment, coupled with climate change that will lead to more and more intense and prolonged droughts, are increasing the risk of seventy-five per cent of the Peninsula becoming desert, according to the Worldwide Fund for Nature (WWF). I belong to a generation that has started to assume that we will soon have to emigrate, given how likely it seems that the dry parts of Spain will become desert during the current century. Actually, it's nothing new to those of us who grew up here; all my childhood, even before I knew it, I dreamed of a future surrounded by northern green. Only when I attempted the move did I see that I had idealized something to which I wasn't suited, and that perhaps the dryness of a place also influences our bond with the land. A

Galician friend of mine, to cope with homesickness when travelling, listens to recordings of rain. Whereas I have a water bottle full of sand from the Sahara, which I still keep so as never to forget what I felt in that desert, and I believe I may now have found my place in a village whose history is marked by supplications for rain. Does the search for water also determine what we consider home to be? As the Joads in *The Grapes of Wrath* say: "This land, this red land, is us; and the flood years and the dust years and the drought years are us. We can't start again."

Something similar occurs, I suppose, with language. They say Galicians have forty different words for rain. We don't have that many in dry Spain, because there's no need for them, but I have added up how many we have for "liquorice" and, if I include "dead man's umbilical"—as coined by my grandfather—as well as the scientific name (*Glycyrrhiza glabra*), I make the total thirty-nine. Growing up as I did surrounded by the succulent, brightly coloured sweets you get at market stalls, I never understood why my grandfather always went around with that shrivelled brown thing sticking out of his mouth. But sucking on that root was his way of keeping thirst at bay, as well as the desire to smoke. *Glycyrrhiza glabra* does not necessarily grow abundantly in cemeteries, but it does in the vicinity of rivers. It seems that liquorice, which in some places is called "Moor's chocolate", originates in North Africa and southern Asia. In ancient times people chewed on it to alleviate respiratory problems, strengthen the muscles and bones, and soften the skin. The Greeks and Romans also used it for another purpose, as recorded by several ancient authors: to combat thirst.

People, according to Hegel, come to resemble the landscape and climate in which they live; if so, that causality would need to be determined, for in order to stay where they are the people of La

Mancha had to form their landscape and their gastronomy to fit their water needs in a region whose toponym could just as well mean "dry land". I come from a place, a landscape and culture largely shaped by water scarcity. There the cereal crops are laid out in geometric figures, a patchwork if seen from above. I come from a place where thousands of years ago my ancestors faced and overcame one of the worst droughts in history.

Far more recently, their descendants saw a stream being covered over with concrete and stopped telling an ancient story. Once, when the stream was still visible, someone discovered a great unidentified slab of something floating in the water upstream, where the women's domain was located: the wash-place. To judge by the general surprise, a whale had appeared. It moved downstream so slowly that the Manchegan whale-spotter had time to go around telling everyone. Men gathered in the village square, shotguns and all, and when the thing finally came within range, they fired. But it wasn't a whale. Rather, it was a donkey's saddlebags. So they used to say in the village, but the story is so steeped in history that nobody can tell if it is legend, joke or hallucination. The whale of the Terrinches stream was also the whale of the Sequillo and Manzanares Rivers, and the latter of these two running through Madrid is how Madrileños got their nickname of "ballenatos" (whale calves). Similar versions of the same story are repeated in other villages in Spain's driest places through which a river or stream flows. As I write, I can see the Guadalope River flowing by. Here, more than 500 kilometres from Terrinches, the story of the whale is also told, a whale that was, in reality, a set of full saddlebags. It even crossed an ocean one day, although it is difficult to know in which direction, because in a Yamaná tale (in Chile), the protagonists also decide to go and hunt it.

✢

According to a UN report, drought has killed 650,000 people in the last fifty years. And an estimated 700 million will be displaced by drought in 2023 alone. Rather than "drought", however, I prefer to think of thirst; sometimes to speak of drought means disregarding abuses, overexploitation and mismanagement of resources. Little is said about the famine that at the time of writing (2023) is ravaging the Horn of Africa, the place from which our ancestors once set out—in search of food or water. Four years of poor rains there and months of none at all have decimated crops, killed livestock, put the lives of millions at risk and forced them to relocate. Even less is said about the causes of this, and when it is discussed, we just say "drought" or "famine". In a text entitled "Drought is Not a Synonym for Famine", Doctors of the World write:

> There is no doubt that a direct link exists between prolonged drought and famine. But it is also true that other factors must be in place for the latter to occur. [...] We must not overlook other causes—such as war, the tyrannical power exercised by many governments over their people, the mismanagement of resources, the inequality implied by the current economic order and the large-scale felling of tropical forests—in order to explain events that could *a priori* be attributed solely to meteorological providence or the misfortune of a country's geography.

Thirst almost never happens in isolation. But in this book it is the protagonist. When I say thirst, I am not just talking about a physiological need far more urgent than any other, but about the absence of water as well, the need to manage and retain it. Our search for water has brought us to the place in which we now find ourselves.

This search has been one of humanity's driving forces. Recent studies have found it in the background of the Romans, who first left their homeland to create an empire, in the fall of the Visigoths and the arrival of the Arabs in the lands in which I write. After forcing our displacement, and tying us to a land, and sending us towards rivers, and making us believe that we can alter nature with no consequences, it made its presence felt in the first war of which there is any record. It also played a more or less significant role in the cognitive, agricultural, scientific, French and industrial revolutions, as well as in the advent of Artificial Intelligence, with which we may end up competing for an increasingly scarce resource. Many people don't even know that a chatbot needs water to function; for one of these to answer ten questions, it *drinks* approximately one litre of water. It is estimated that as AI's use becomes more widespread, its water consumption could increase fivefold. But at the same time, AI is helping to improve water conditions in some refugee camps.

The major revolutions that have led us to dam rivers and dry up aquifers to meet our water, food and electricity needs have already had a knock-on effect on the rain that falls and our not dying of cold or heat. Our search for water is capable of altering the movement of the Earth, as the Three Gorges dam in China—home to the world's largest hydroelectric power station—has already done. Experts say that we will be unaffected, that the huge accumulation of water in one place is lengthening the days by only 0.06 microseconds and that the massive extraction of water from aquifers has shifted the Earth's axis of rotation by only 80 centimetres in a decade; these things, after all, are constantly changing. But if long-term climate change is in large part down to precisely such variations, how can we be sure that they will not influence a future climate that we humans will not be around to experience?

I also choose to focus on thirst rather than "drought" in order to give water scarcity its historical dues *without* the excesses of environmental determinism, which relegates humans to the role of weather-controlled puppets. The agricultural revolution made our ancestors dependent on rainfall as never before, and it was then, too, that a considerable part of our species began to leave its mark on the planet as never before. But drought was only one of the causes. Drought did not in itself provoke revolutions, though it did often lead to famines and epidemics that collided with despotism.

This book is neither memoir nor essay, but a hybrid. Using childhood memories as a jumping-off point, I have tried to understand why wine, bread, oil and pork are such staples in La Mancha. Where we come from and why we left. Why we stopped moving on and started praying to rain gods. Why so many hunger strikes were preceded by drought years. Why a farmer who apparently lived in Madrid 900 years ago is such a presence in my village today. And how we have tried to manage rainfall and retain water through both traditional and scientific methods.

In Part I, some stories from my family lead me to trace humanity's journey, namely from Africa to the Iberian peninsula, where all those I can name among my ancestors have lived. Humanity's thirst has been a more powerful migratory driver than love has, putting us perpetually on the move, until a point came when we stopped—more or less—and began to cultivate the soil, while also beginning to gaze skyward in hope of rain. But before reaching prehistoric La Mancha, where Europe's first hydraulic society may have arisen, we will stop off in the Fertile Crescent. This was one of the places where human beings discovered the ability to cultivate the soil before settling there to await the rains, and eventually learning irrigation skills too. We will see how various periods of cooling and drying during the early Holocene displaced large numbers of tribes, who then slowly began establishing themselves in the environs of the largest rivers of the day. We will also see how, because of thirst, climate refugees came to found new civilizations, as well as retaining languages, which until then had always come and gone. Cities, kingdoms and even the first empire perished, largely as a result of one of the most severe and prolonged episodes of drought in places as far apart as Mesopotamia, the Indus Valley and what is now Peru. In the meantime, the prehistoric Manchegans, among whom the Yamnaya were already living, managed to survive by extracting groundwater, until flooding came along.

Part II begins with what may have been the first-ever readings of the constellations, which preceded rain gods: animalic and then anthropoid gods, and finally deified humans. People's search for water led not only to faith and supplication, but also punishment of those who managed or claimed to manage the rain—be it a god-king, shaman, witch, saint or meteorologist. We conclude with a focus on traditional techniques for monitoring rainfall, on the sciences that have gone on to supplant them, and on the people who began studying the sky and were the first to give names to different clouds, to predict storms, measure rainfall intensity and droplet size.

Finally, looking into family memories and parish records, I will discover a family tree in which I once again come upon thirst and a "severe drought" that may not have been so severe, or not enough to cause a famine. I also go to the new Riaño to find out what toy the roof-mounting child chose to take with him before his house was flooded—for the drowned and the thirsty, equally exiled by thirst, have a shared fate and shared pain. What shall we call them from now on if they (we) are to become more and more numerous?

At all times I have tried to ensure that this story goes beyond white European men simply eclipsing one another in turn; that it not be limited to humans, because the camel, the oryx and the goose have also found ways to fight thirst; that it transcend scientific and urban elites, because the popular wisdom of rural women and men is not only compatible with science but is often its starting point, given that the teachings of so many proverbs have come to be scientifically proven.

This journey has led me to debates past and present in anthropology, palaeontology, climatology, genetics and above all in archaeology. And it has also taken me, figuratively speaking, to other drought-ridden places in the world that allow glimpses of part of a whole and that connect back to the starting point. It has taken an effort for me, a journalist and social and cultural anthropologist by training, to understand and convey in an accessible way ideas and concepts previously unfamiliar to me. For this reason, and because I wanted to make it easier to read for those who are not familiar with certain disciplines, I have omitted some names, facts and dates. Many of the scientists and popularizers without whom I would not have been able to write these pages are mentioned in the bibliography and, in some cases, also in the acknowledgements, because they helped me resolve things I was uncertain about. Any mistakes however are mine alone, and that I have done little more here than share my amazement as I have gone about seeking answers while coming across thirst in the most remote places, and crucial turning points for humanity. I have often had to curb my enthusiasm because, wherever I looked, thirst lurked.

I

THE JOURNEY OF THIRST

1

Heavenly pork

> No trace of humidity, no memory of water came to save us from the set of thirsty reflections.
>
> ELENA GARRO,
> *RECOLLECTIONS OF THINGS TO COME*

The house in which I first experienced our thirst for water is a time capsule. The water tank, the jug, the candles we would use during power cuts, and a shepherd's canteen—made from a gourd by my great-grandfather Pedro—are all still there. That great-grandfather cut open the gourd, scraped out its flesh and seeds, cured it and inserted a cork stopper into the hole he'd made. He kept it on a cord about his neck, drinking from it on his long days out in the fields. I find it curious that the "cal-" of "calabash" goes back to words that mean "shelter", "house", and "shell".

Although not easy to grow in dry climates, gourds have always been used to combat thirst, as much in the drought-stifled Yucatán before the collapse of the Mayan Empire as in La Mancha. Jaguar Paw, the main character in the Mel Gibson movie *Apocalypto*, has just such a gourd for his canteen, known as a *guaje* in Mexico. *Lagenaria siceraria* was being cultivated there almost 10,000 years ago. In fact, they were one of the first plants ever cultivated by human beings, and are described in Navajo myth as *the* first. Their woody texture, bland flavour and toughness hardly make them appetizing, but so useful were they as receptacles that they were very popular to grow. Given their capacity to float on the sea for two years without losing any seeds, it is thought that gourds could have made the journey from the Americas unaided. They attained such importance that 2,000 years ago they were buried with dead people in tombs as far apart as Peru and Egypt.

In my village there was a man who gave them another use, transforming them into art. I came across Juan the miller in the street one day, penknife in hand. He was splitting a dried gourd. Although in La Mancha gourds grow with no other aspirations than to become canteens, my grandmother's neighbour was making a lamp out of this one. And it wasn't the first. His house made for an eccentric museum, full to the rafters with repurposed gourds. That same day Juan told me that he was one of the many men from the village who had been employed as extras in *Spartacus*. They were the slaves who fought under Kirk Douglas's Spartacus, the Thracian who rose up against the Roman Republic, sweeping the people along with him, and whom a recent TV series has turned into a "rainmaker". Among the thousands in the most well-known scene in the Stanley Kubrick film were many members of the Spanish army, Juan the miller among them. The director managed to shoot the scene on the acceptance of a condition imposed by Franco: he could have the soldiers, but only if they were not shown dead on screen. And so along they went, in exchange for a sandwich and a few pesetas.

Juan didn't just talk to me about blockbuster movies and gourds that light up the darkness. He shared a story that I found extremely funny, though for him it had been a matter of life and death. He boasted of always having had an iron constitution and only once being admitted to hospital. One day, the nurses brought him a yoghurt, and Juan complained: "But where are the bones in this?" Although it wasn't the bones themselves whose absence he was lamenting. After giving the idea some thought and finding himself close to despair, he decided to undertake a quixotic feat and made a run for it. The nurses chased him down the corridor, but he managed to escape, and thereby achieved his goal: returning to eat a slice of bacon in the comfort of his own home.

As well as Juan's story, during that trip home I discovered that one of my former neighbours had an exceptional command of culinary alchemy. I heard her tell how, every day after siesta, she made her son a "vegetarian sandwich". It had cheese and bacon in it, but she never omitted a little lettuce and tomato. Who hasn't experienced the Iberian magic of tuna salads, which qualify as "vegetarian" because of the tiny slivers of lettuce and tomato? I quickly came to understand that my family's pork obsession was more than a matter of personal taste, and that, if other Terrinches men agreed to take part in the filming of *Spartacus* in exchange for a sandwich, there would have been meat involved, and it would have been nothing lean.

There was a corner of my grandmother Araceli's house that she guarded so jealously that it was always off limits to me. It was a pantry smelling of rancid pork and, since I was only a child then, I can admit without shame that I once locked her in there and went out into the street, proud to have left her alone with a love she was so protective of. The memory is infused with the smell of oranges, so perhaps I snacked placidly on her doorstep while she pleaded to be let out. As for my grandmother Francisca, she could forego anything in life except a little bacon on bread at dinner time. She embodies the ordeal of any vegetarian or coeliac granddaughter. My being among the latter, she once offered me chorizo to dip in my milk because she couldn't get any gluten-free pastries at the village bakery. I declined but she, undeterred, offered me ham instead. She often advises me to reduce my intake of vegetables, considering them fit only for livestock. Among her disparaging comments about certain foodstuffs, which would frighten any nutritionist, are phrases along the lines of Juan's, such as "there's no kidney where that comes from". I don't know if there's a single house in La Mancha without some cold meats in the fridge to serve as dessert for anyone who isn't full. Then again, that never happens. Marvin Harris says in his book *Good to Eat* that villages and groups studied by anthropologists have shown an unfaltering obsession with meat because of the way it fosters social ties.

It matters little whether the famously mysterious "duelos y quebrantos" eaten by Don Quixote on Saturdays (variously translated as "scrapings", "bubble and squeak" and "eggs and bacon") is a genuine Manchegan dish or an invention on Cervantes's part, because it includes one star ingredient that makes it very much real; save for the lentils he eats on Fridays and the pigeons on Sundays, bacon is the ingredient most frequently featured in *Don Quixote*. On the other days of the week, his stew would have had meat in it, and his *salpicón*, which nowadays we would think of as a seafood salad, would instead have included leftovers re-fried with onion and bacon. That is, fat upon fat, in a land where we even have a dish that is just fried breadcrumbs. Later in the book we see a sort of Spanish pease pudding that includes bacon, and a *morteruelo* pasta dish also featuring bacon. Don Quixote's pottage, similar to *morteruelo*, is finished with the star ingredient. It is not clear whether in *Don*

Quixote hunks of bacon are scattered over the crumbs or the pease pudding, but in today's La Mancha they are a must; pease pudding, with which our grandparents kept hunger at bay in years of scarcity, is now a favourite dish among their grandchildren.

Cervantes wrote *Don Quixote* at the height of the Little Ice Age. For 500 years or so, cold and drought dominated much of the world. This might be why it rains just twice in the novel (plus it takes place in the epicentre of dry Spain), and two of the best chapters have the protagonists' thirst, and a prayer *pro pluviam*, as their starting point. Although we will return to the world of *Don Quixote*, it is worth noting here that pigs were then commonly called "swine", and Spain was deeply divided between porcophiles and porcophobes, a fact that usually denoted whether they were "old" Christians or "new" (that is, converts from religions like Judaism and Islam). They were something like today's *concebollistas* and *sincebollistas* in Spain, famously at loggerheads over whether tortilla should be cooked *with* or *without* onion. In the Golden Age, people also tended to beg pardon if they accidentally let slip the name of an animal that was adored and reviled in equal measure.

There has recently been the very revealing case of a researcher who came up with his own bold pizza topping: instead of *fuet* (a kind of spicy sausage containing bacon), it had melon. This was aimed at reducing bacon consumption for the sake of people's arteries rather than provoking my grandmother, but she was still unimpressed: "That sounds like poison," she said. "Don't you go anywhere near it till all the facts are out." We are talking about an eminently porcophile place where people enjoy melon, especially if it comes from Tomelloso, but less so any sleight of hand that means missing out on bacon. People like my grandmother have their reasons for eating what they eat, for prioritizing that which "comes from something with a kidney". Post-war privations undeniably had an impact, but the reasons have a longer history still.

When La Mancha ceased to be a demographic desert in the twelfth century, the people who repopulated the area made olive oil the centre and mainstay of their gastronomy. It was a good place for olive growing. After all, the wild olive tree appeared in Mediterranean forests some 150,000 years ago, in conditions even more arid than today's. But when the population mushroomed, oil became scarce and for a time had to be brought in from Andalusia. There was land aplenty for drought-resistant olive groves, and from the eighteenth century onwards these became more widespread, with almost every dish from then on containing our very own holy trinity: bread, oil and some pork cut. With this foundation, especially as olive groves expanded, the cuisine of La Mancha became increasingly sophisticated. The first Manchegan folk song I learned to dance to sums up the building blocks of my homeland's cooking: "To Manchegan La Mancha,/ where the wine and the bread/ the oil and the pork/ are plentiful." Vineyards, olive trees, pigs and cereal. All of them, to varying degrees, proliferated because they could withstand the climate, and in turn reshaped the landscape.

Although potatoes were already being grown in La Mancha at that time, they were banned in parts of Spain and Europe for centuries because, apparently, some Spaniard had the idea of eating them raw, skin, soil and all, prompting word of their inedibility to spread. My ancestors went without fish for centuries, the sea being so far away; their enthusiasm for meat was passed down to both my grandmothers. But then there was the coming of the railway and, with it, the blessed Basques, who furnished new dishes with their *bacalao* or salt cod. Then along came the famous *atascaburras* (salt cod with mashed potatoes and garlic) and *tiznao* (salt cod with spring onions), bringing joy to the palate and to gastronomic nomenclature alike.

Indirectly, in discussions of alimentary taboos and the reasons behind them, the anthropologist Marvin Harris assigns an interesting role to the search for water.

Although pigs were domesticated in the Near East at a time when the vast forests there had not been entirely replaced by grasslands, bearing in mind all the shade and water they require and the fact that they provide neither milk nor clothing, they were a considerable drain on the very resources required by humans, especially during droughts. In explaining the biblical and koranic condemnation of pigs, and making connections to the sacredness of cows in India, Harris is in fact discussing the same thing: how we include or discard certain foods in our cuisine based on what is possible in arid conditions. The Israelites and Muhammed's original followers alike lived in desert places, and one can't help but wonder whether a farmer in India would refuse cow if he didn't depend on its calves to plough the desiccated land, or whether a Muslim would happily eat pork if her ancestors had not been competing with that animal for resources in semi-desert places. But a paradox is also possible: that a pork-eating society could exist in arid lands and the religion of that place prohibit meat only at certain times, just as, for reasons of identity, pigs are still taboo among Jews and Muslims who no longer live in the Near East.

Does it therefore make sense for pigs to be a crucial element in a gastronomy rooted in an arid land? Perhaps, perhaps not. Although at first glance not the most obvious adaptive choice, it fulfilled the social function we have already seen in Cervantes's day, as well as being one of the most affordable sources of animal protein. But the importance of an animal that can be eaten "trotters and all" is clear when it can provide a family with food throughout the year. Not to mention that it can co-exist, to a large degree, with goats and sheep and drought-resistant crops. But there may be another, less obvious reason. I usually associate high-fat diets with cold climates because I too sometimes forget about own my thirst, but perhaps the Manchegan obsession with pork is related, in some way, to the fact that camels accumulate fat inside their humps.

The predecessors of camels migrated from the Americas to Eurasia and Africa during the ice ages. While they died out in the places they were originally from, they adapted to the extreme conditions of their new homes. In Africa, and along with *Australopithecus*—our early hominin ancestor—they developed an astonishing capacity to accumulate fats, which enabled them to survive in very dry and hostile environments. Unlike other macronutrients, fat can be accumulated in the body without the need of water. Additionally, when metabolized, it converts not only into energy, but particularly into water. So the water contained inside camels' humps is a fable, but not entirely: just as they have three eyelids and the ability to close their nostrils when a sandstorm hits, it is fat that they accumulate in their humps, but this in a parched body is *effectively* water (or the metabolic water produced by lipids when oxidized). This, together with their huge internal water reserves (they can drink up to 114 litres at a time) and the ability to defecate waterless droppings, allows them to survive for days, weeks and months in the desert with neither food nor water.

Similarly, there are examples of animal and plant behaviours that speak of nurture as well as evolutionary nature. While the Australian water-holding frog and certain desert tortoises store water across their entire bodies and can live without it for up to five years, koala bears drink no water at all and make do with the liquid they extract from eucalyptus leaves, while the Namibian desert beetle draws water from fog. The American spadefoot toad is fascinating in this respect: in times of drought it buries itself in the soil for months to retain water, only emerging when it senses approaching rain. The African lungfish, for its part, can survive without water despite being an aquatic animal; when the shallow swamps and marshes in which it lives dry out, it burrows down into the hardened mud and covers itself in saliva as a way of retaining moisture. It then sleeps until awoken by the rain.

Some desert plants have also adapted to store the tiny amounts of rain they receive. The saxaul shrub, a traditional desert remedy for memory loss, holds salt in its leaves to improve water absorption and thereby survive conditions in the Gobi. As if that were not enough, it has also succeeded in producing energy through its branches to retain water, which it stores in its bark and roots. The difficulties of grazing in the Gobi, which is increasingly often snowbound, have driven more and more people to the Mongolian capital of Ulaanbaatar, while the hardy saxaul lives on, making do with the barest drop of water.

Humans, unlike other living beings, adapt *culturally*; to a large extent, culture, which operates at a faster pace, has made us independent of our environment. The Spanish anthropologist Juan Luis Arsuaga, glossing the work of German evolutionary biologist Ernst Mayr, puts it like this: "Our organs do not need to adapt to different ecosystems, because we make tools to do the job for us. Whether a spade to dig with or a canteen to drink from, these, to all intents and purposes, can be considered artificial organs, prostheses." In other words, my great-grandfather did not need a water storage tank inside his body because thousands of years ago in Mesoamerica someone began growing gourds, and he turned these into the receptacles he needed. Neither is it something my grandmother or her neighbour have ever had need of because they belong to a people whose gastronomy is entirely determined by water scarcity. But if one of its ingredients, pork, has made life easier in such a dry environment it is precisely because their ancestors did adapt their bodies, and they inherited these adaptations.

Though we do not need new anatomical and metabolic adaptations as much as other members of the animal kingdom do, it doesn't mean we don't adapt at all. There is ample proof that humans have developed superior lung capacity when living at altitudes where the air is thinner—like in Bolivia and Tibet. Similarly, islanders in southeast Asia are able to hold their breath for longer and have better underwater vision, while members of livestock-farming cultures have been found to be more lactose-tolerant. In addition, some Europeans have larger noses as an inheritance from *Homo neanderthalensis*, which would seem to have made it easier for them to live in cold, dry places—as we shall see later on.

In certain cases, other animal adaptations, and those of plants, have proved advantageous to humans, who have learned to use and sometimes abuse them. The San people of the Kalahari, for instance, having observed that thirsty monkeys can locate water in the desert, capture baboons, deprive them of water, and then follow their lead once they have freed them. In La Mancha and the coldest parts of Norway, just as in the Kalahari, adapting by way of cultural innovation is part of what has enabled humans to go on living in such harsh places. In this sense, even if they have taken different paths, it is for similar reasons that camels have humps and my grandmother can't eat a meal that doesn't include pork. This also goes for another grandmother—one as much yours as mine—from whom we inherited an adaptation that made pork an advantage in arid places. But first, for a brief and admittedly somewhat simplistic tour of Lucy, the grandmother of all mankind.

In the very beginning, there could be no thirst, because there was neither water nor life on Earth. According to one of the most widely accepted hypotheses, our planet started out as a disc-shaped amalgam, a fragment of the ruins of a collapse that took place within an expanding cosmos. The disc became a not-quite-ball—slightly pinched or flattened at the poles—wheeling around itself and around a giant star, spinning top-like. Against all odds, and by pure chance, it ended up at the exact point, neither too far from nor too close to the sun, at which water could be retained in a liquid state, and thus the conditions for life could begin to emerge. But this was still a long way off, because a burning, dry, inert ball of rock under constant bombardment by comets was not habitable. Unsupervised works began that would leave a house almost ready to be occupied, but one not without its quirks. In the early twentieth century, astronomer Milutin Milanković wanted to understand long-term climate change and spent thirty years studying these quirks, which he termed "defects". He set out what became known as "Milanković cycles", although at the time his work was not taken very seriously. Later paleoclimatic studies have revealed that the Earth has experienced changes in climate on a periodic basis strikingly similar to the one established by the Serb. These cycles do not occur independent of other factors, but on the basis of solar radiation—which hits the equator more than the poles. No person will ever live through these phases in full, which depend on orbital variations caused by three things.

Two of these have to do with the imaginary axis about which the Earth rotates. On the one hand, the fact that it is not perpendicular to the plane of the Earth's orbit is what gives rise to the seasons.

However, the degree of tilt (or obliquity) is not constant, altering in cycles of 41,000 years, which in turn alter the amount of solar energy we receive. On the other hand, the precession of the equinoxes is due to variations in the amount of sunlight received at the beginning of a given season, which builds up and results in milder or harsher seasons. Because the Earth spins at an angle, the motion of the axis forms an imaginary cone that sometimes widens, increasing the roll or pitch of the Earth every 26,000 years. Particularly interesting for our purposes here is the moment when the axis is at its *most* tilted, given that this leads to an expansion of both deserts and ice caps, while shrinking the temperate zones.

The latter depends on the proximity of our ruling star, which increases and decreases in cycles of 100,000 years—which, as it turns out, is the length of the ice ages experienced by our planet—and of 400,000 years. The Earth does not move in perfect circles around the sun; its trajectory is rather an ellipse. Its path is also modified, with a slight lengthening of its orbit, by Jupiter's and Saturn's gravitational pulls. These variations cause it to receive more or less solar radiation.

Ultimately, these large-scale changes in climate have several contributing factors, including: the Earth's oblate shape (being pinched or flattened at the poles); the varying degrees of its imaginary axis; the fact that its motion around the sun is not perfectly circular; and the sun not being located at the exact centre of its orbit. Although the point at which the Earth randomly took up position (the so-called "point of stellar habitability") allowed life to emerge, our arrival came at a cost. It is precisely this location that complicates our existence. Climate change is the price we pay.

A great deal of time passed between our planet coming into existence and the birth of the first living being. It was millions of years

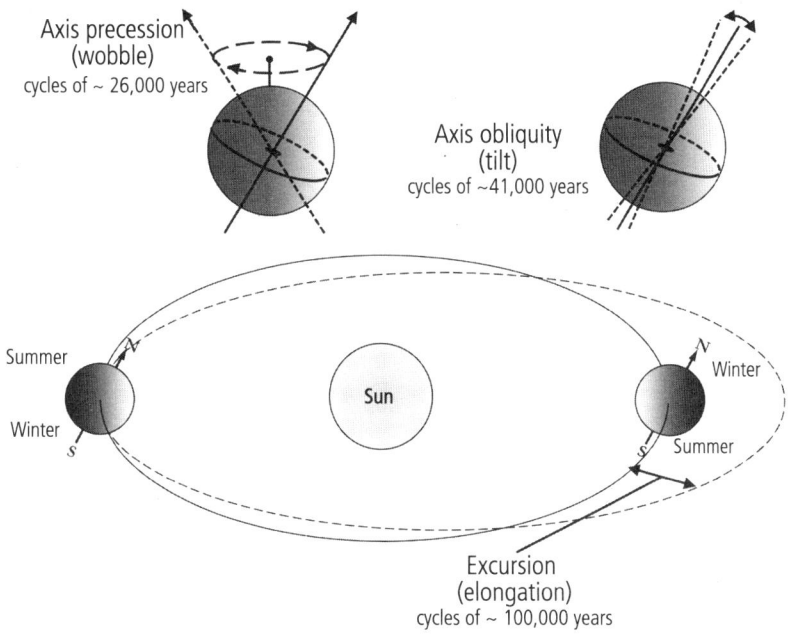

Milanković cycles: variations in the Earth's orbit and axial tilt that affect our climate.

before the emergence of LUCA, a bacterium from which all living things on Earth are descended. LUCA is often said to be the first, but neither was it the first, nor was it alone: as its name suggests, it is simply the *last* universal common ancestor. Hence we are related, to a greater or lesser degree, to cats, geraniums, mushrooms, flies, chimpanzees and kākāpōs. To bring some order to the boxes inside which the Earth stores living things, biologists have devised a hierarchy that allows them to group us into a kind of schematic tree, which they call a "taxonomic category". Following this classification, and considering only the most relevant branches in our story here, we can consider ourselves eukaryotes (living beings), animals (kingdom), chordates (phylum), mammals (class), primates (order), *Hominidae* (family), *Homo* (genus), *sapiens* (species), *sapiens*

sapiens (subspecies). There are intermediate steps between these categories, among which we are interested in highlighting some that proceed from our family to our species, such as hominins (tribe, which includes humans, our extinct ancestors, and chimpanzees and bonobos) and hominids (subtribe of upright, bipedal primates of which today only *Homo sapiens* remains). Scientists have recently tweaked this order in ways that could send this book on a significant detour; suffice it to say that I will call both present-day humans and their extinct ancestors—the ones that moved away from the chimpanzee evolutionary line—hominins.

Tempting as it might be to believe that thirst is as old as life, it took LUCA's descendants a long time to become multicellular, much longer to develop a backbone and even longer to emerge from the water. Could thirst have existed during the 3.8 billion years we spent in the water? Or did it emerge when we first ventured onto dry land, initially as reptiles and then as mammals? Well, as the Armenian saying goes, "When you fall in the water, you don't worry about the rain." Mammals have been around for some 200 million years, and we primates appeared in Europe and North America (these continents were not where they are now) about seventy million years ago. They initially fed on insects. But one day plants appeared, flowers and fruits sprouted, and they also began to eat these with gusto. And so we continue to do, those of us hominids who appeared some forty million years ago in Africa, when the Earth had already entered the cooling-drying arid phase in which we still find ourselves. Those ancestors of ours stayed up in the trees and increasingly focused on fruit, but at some point adopted bipedal locomotion. At first it seemed a futile strategy, until fruit and trees became more scarce and they were forced onto the savannah to forage. We did not fully diverge from the chimpanzee branch until about seventy million years ago (the divergence began far earlier), and it wasn't until two to three million years ago—when cold and aridity reached a new peak and the search for water once more became a driving force in East Africa—that we could be considered properly "human". In trying to return home whenever life becomes challenging, it is possible that as humans we also emerged alongside a river, the Omo. The palaeontologist Yves Coppens could not resist the pun, and called this new leap—one crucially affected by a crisis of aridity, which gave rise to *Homo habilis* and may have been our moment of origin—the "(H)Omo event".

Other species would follow, but to stay with *Homo ergaster*: these ancestors of modern humanity tripled their brain size compared to that of *Australopithecus*, and with this change came the perfection of tool-making, the invention of the hand-axe—prehistory's Swiss Army knife—and the first forays beyond Africa. They were also perhaps the first to produce fire, not just harness or maintain it, as well as to engage in hunting, gathering and scavenging. Those who left Africa went through several significant evolutionary steps, becoming the *Homo neanderthalensis* of Europe and *Homo denisova* of Asia. Those who remained in Africa evolved and gave rise, 200,000 years ago, *to Homo sapiens*. These, in turn, evolved into *Homo sapiens sapiens* (anatomically modern humans—that's us). The key dates in our evolution are remarkably in keeping with prolonged changes in climate that in most cases involved cold and aridity, as happened when we were on the brink of extinction shortly before we once more ventured beyond African shores. Are we, then, children of climate change, crises, cold or thirst? A little bit of each, is the likely answer.

The fact that today we are *sapiens sapiens*, doubly "wise", does not imply an intelligence so superior to that of our nearest ancestors, as was once widely thought. Since we were also given this name by men who belonged to a culture that still believed women came from the rib of their counterparts, many people got stuck on the idea of us as apes, and embellished it with claims that we came from monkeys; as if a chimpanzee had suddenly given birth to the first human being. Women in the nineteenth century had to accept that they came from a rib that in turn came from mud, but they could not accept that they were also the cousins of chimps and gorillas—which were not in fact less evolved than we are, only differently so, in context and direction. They did not yet know about LUCA! Although Darwin spoke as early as 1859 of an ancestor common to all living things that would have been born in a "warm little pond", it took a century for the genetic code to be deciphered, which was when its universality became known.

In conclusion, we are apes and we come from bacteria, whether we like it or not. On top of that, we are also insignificant, a dash of water that longs for water, made of stardust, on a planet that, although it seems huge, is itself nothing more than a tiny ball circling a star in a galaxy. And the Milky Way, as lovely as it may sound, is just another speck in a cluster of galaxies, which in turn is just another speck in the universe. It is fairly bold to think that the universe conspires in our favour, but it is also comforting to believe that it sends us signals when we lose our way. We couldn't even decide how we evolved because evolution works on its own and doesn't make decisions, doesn't ask for our opinion and doesn't follow a route with a fixed destination. Nobody is really at the wheel. Humans are tiny, but for all that, fascinating, because I could not write this summary without the questions posed by a series of dazzling minds that have gone before, and the reams of pages they devoted to the answers. I want to stress that these are all *possible* answers, because in science there are neither dogmas nor certainties. Religion and science are not so incompatible in essence: they often ask the same question everyone asks once they become aware of death and are able to string together more or less complex sequences of words: where do we come from and where are we going? Only the answers differ: science tries to explain how things might be and expects critique, while religion states how things are. Both have a place in this book because religion, which came first, is also studied in the social sciences and shows the search for water to be one of our very earliest preoccupations.

For now, we will stay with science and, in the aforementioned hierarchy, with hominins, as we travel back to a time when neither the genus nor the species to which we belong existed. Now to meet the grandmother of mankind.

Long ago in a place called Hadar, in the Afar region of Ethiopia, there was a lake. Nowadays there is a desert. It hardly ever rains there, but when it does, the water falls so suddenly that the treeless land struggles to absorb it. And this means the most unexpected things can be found in the dried-up lakebed.

Donald Johanson went to Hadar in 1974 as chief palaeoanthropologist on an international, multidisciplinary research trip. He was a superstitious man with clear dreams and goals. Some time before he had found a knee joint that went on to form the first evidence of bipedalism among hominids. His friend Owen Lovejoy, an anatomist and expert in locomotion, asked him to return to Africa and bring him back the complete body, and so he went.

On 24 November 1974, Johanson had resolved to spend the morning on a backlog of paperwork. But a hunch prompted him to go and talk to Tom Gray, a doctoral student studying the historical co-existence of animals and plants in the area and how this related to the climate. Johanson suggested a walk to find a site that had not yet been located on a map. After hours of scouring a ravine, and with temperatures soaring close to 43°C, the palaeoanthropologist wanted to take one last look, because he had woken up that day with the firm conviction that luck was on his side. Just before returning to the Land Rover, he spotted a piece of bone.

> "That's a bit of a hominid arm," I said.
> "Can't be. It's too small. Has to be a monkey of some kind."
> We knelt to examine it.
> "Much too small," said Gray again.
> I shook my head. "Hominid."
> "What make you so sure?" he said.
> "That piece right next to your hand. That's hominid too." [...]
> "Look at that," said Gray. "Ribs."
> "I can't believe it," I said. "I just can't believe it."

Then they saw another bone fragment. And another, and another. In the end, the fragments constituted forty per cent of a single skeleton.

Whoever the person had been, they had died there, at the bottom of a lake that was now part of a desert. Hidden under sand and mud for more than three million years, one day the rain came and exposed the bones once more. That evening they celebrated the find back at the camp with other members of the expedition, the leaders of which included the Frenchmen Yves Coppens and Maurice Taieb. They spent the evening dancing and singing along to a Beatles song playing on a loop on the tape recorder. When "Lucy in the Sky with Diamonds" started for the umpteenth time, one of them decided that the grandmother of mankind should be called Lucy.

HEAVENLY PORK

In the late nineteenth century, the African leader Makapan and his tribe were besieged by Boers in a cave. They soon discovered that this place could be the saving of them, and used it as a refuge and fort against their attackers. They banked up enormous boulders to block up the narrow entrance. They held out for weeks, before being forced out; Johanson, in his book about Lucy, says "they were eventually driven out by thirst". Some 2,000 people were massacred as they attempted to escape, and 1,000 more were killed inside.

At the time of the massacre after which the Makapansgat Cave would later be named, it seemed scarcely credible that humanity could have originated in Africa, even though Charles Darwin had speculated as much almost half a century earlier and Thomas Henry Huxley was by now shouting it from the rooftops. Darwin's *On the Origin of Species* changed what humans saw when they looked in the mirror, but not everyone wanted to look. Evolution was not a completely new concept, nor did Darwin coin the term, but his great achievement was to identify what drove it—natural selection—which to this day has no convincing alternative explanation. He initially avoided being too direct about humans, perhaps because he had an ace up his sleeve, as he himself seemed to imply and as he explained shortly afterwards in *The Descent of Man*. Doubtless he was aware of the likely repercussions if he came out with it all at once. His peers largely still believed in God the creator, and themselves as one of His creations. How to break it to them that perhaps there was a rather different cause?

While Darwin was hiding away in his home village after the famously tantalizing sentence near the end of *On the Origin of Species*—"Light will be thrown on the origin of man and his history"—Huxley was so bold as to draw direct links to gorillas and chimpanzees, deducing that the origin of mankind must be in the places still inhabited by them. For Darwin, we do not come from apes, as became commonplace; we *are* apes. If fossils of a common ancestor existed, they had to be in Africa. It didn't take long for the objections to come pouring in. Few wanted to accept that such an ancestor had resided outside Europe or Asia, far less acknowledge that the cradle might have been in Africa. One of the clues lay in the Makapansgat Cave.

There is no way for me to know what my great-grandparents would have thought when Darwin's adulterated ideas began to enter the culture. They were illiterate, but such was the furore that I imagine they caught wind of them, however distorted. Something along the lines of: "There's an Englishman who says we come from monkeys! *He's* the son of a monkey, I say!" The scorn for Darwin took many forms, even providing inspiration in Spain for a drink branded Monkey Anisette, the image on the label of which is believed to be a caricature of the scientist.

Those who accepted evolution had the chance to assert the ancient nature of their geographically proximate ancestors, but their misunderstanding of the theory sent them down the wrong track. At that time only a few Neanderthal fossils had been found, and scientists did not fully accept these as belonging to a possible ancestor or relative, rather dismissing them as deriving from some deformed ape. *Barbarian, Cossack, cur, old Dutchman* (said with contempt by a German), *shaggy back, freak, Celt of poor mental organization*; these are not alternative words to a playground song, but just some of the pleasantries levelled at *Homo neanderthalensis* by scientists as increasing numbers of remains were uncovered on the European mainland. Some time later, Cro-Magnon Man was found in southern France. Since he was a *Homo sapiens*, and therefore felt to be closer to our kind, people were more indulgent than with *Homo neanderthalensis*. But the general idea was still of a slow-witted, drooling caveman, albeit one less crude than *Homo neanderthalensis*. Perhaps it was nothing to be proud of, but neither was it much to be ashamed of. There was a widespread view that the origin of humanity had to lie in Europe some 100,000 years earlier. But then the focus moved further afield.

Between 1887 and 1895, Eugène Dubois went in search of the missing link, first on the island of Sumatra and then on Java, both of which were home to orangutans. Originally in Southeast Asia as a medic in the Dutch army, after contracting malaria he was sent to join the reserves in Java. Only then was he able to devote himself full-time to his true interest. He found the half-million-year-old Java Man, and the origin of mankind was consequently both sent further back in time and relocated. Heidelberg Man was found in Germany, and Peking Man in China, and they were all grouped together under the title of *Homo erectus*. Fossils were found in places other than Europe and Asia as well, but almost nobody went looking in Africa, where their common ancestor *Homo ergaster* had in fact once existed. In Johanson's explanation, the logic went that "[h]ominids are descended from apes, apes live in tropical forests, and for millions of years there have been no tropical forests in South Africa." But this neglected the possibility that, if they did exist millions of years ago, the remains of their inhabitants might still be there.

One day in 1924, the South African anthropologist Josephine Salmons, at the time an anatomy student with a passion for fossils, had a hunch that a skull she had seen mounted on somebody's chimney place was not that of a primate so removed from us as it might have seemed, and she took the idea to the physician and palaeontologist Robert Broom. He rejected the possibility, and she turned to Raymond Dart, her anatomy professor. Dart was unable to analyse the crate of fossils as quickly as he would have liked, as when it arrived he was about to go out to a wedding. The groom called for him just as Dart thought he could discern a human skull peeking out from inside. When the wedding was finally over, he found among the fossils a piece of skull which he thought might indeed be that of a baboon. But he began to change his mind as he proceeded to fit several other pieces together. It suddenly seemed to him that the specimen in question was anthropoid, and also that it had walked upright. He published a paper in *Nature* that completely overturned established thinking on humanity's origins. Thanks to Salmons, he had just found the Taung Child, the first evidence of upright, two-legged walking. He called it *Australopithecus africanus* and announced it as the missing link. His publication was so hasty that anthropologists doubted Dart and turned their backs on him. Broom was his only supporter.

Some years later, when Dart had been unable to find anything further to say about the Taung Child, he attended the lavish presentation of Peking Man in London, taking his specimen with him. His friends took him out to dinner to cheer him up and, thinking this would not be an ideal place for the skull, he gave it to his wife Dora, who then left it in a taxi. Reportedly, when the police received the call from the taxi driver, who had been driving the skull around all night, they initially suspected murder. This might have triggered the frustration that drove Dart to desist as a fossil hunter. Nevertheless, in the middle of the century, he travelled to the cave that had been the site of Makapan's last stand. By then, the South African government had turned it into a historical monument. There, twenty years after the London trip, Dart rediscovered his enthusiasm and decided to go back to palaeontology. In the cave of the infamous massacre he uncovered a similar, but much older, story. Of the forty-two baboon skulls he found, more than half had been partially crushed. On the basis of broken fossils and the remains of apparently bipedal pelvises, he came to a conclusion that would create new problems for him. Dart, who had discovered *Australopithecus*, ended up depicting them as bloodthirsty killers, violent murderers and eaters of baboons, which in turn fuelled the idea that humanity was inherently violent. Although the fossils he found had largely been broken down by hyenas and geological processes, Dart's ideas influenced many novelists and also inspired Stanley Kubrick's *2001: A Space Odyssey*. The "killer monkey" hypothesis boiled down to the idea that the urge to kill, whether by hunting or murder, was what led our ancestors to climb down from trees and start fashioning weapons; only later would they stand up on two feet, which, in turn, would have allowed them to wield and hurl weapons. In other words, according to this controversial theory, a murderous, hunting and even cannibalistic urge is what made us human. Needless to say, if we really were killer apes, we would have had plenty of time to become extinct by now.

The year the killer-monkey theory hit the headlines also saw the commencement of a great hoax, one that persisted for forty years. Other countries now had their Cro-Magnons and their Neanderthals, and the English refused to be left behind, however contemptible those creatures might have seemed to them. So, as if by magic, in 1912 Piltdown Man appeared. Perhaps because the search for our origins began in Europe, and despite the initial findings being such a disappointment to so many, there has been a curious drive ever since to show, contrary to the view of most scientists, that we come from Europe and not Africa. Meaning that even in recent years it has not been uncommon to see books claiming a European origin for *Homo sapiens*. How could any reasonable person think that we, the truly wise, do not come from here originally?

Piltdown Man was proudly presented to the world as the missing link, and the first Englishman to boot. Darwin's ancestor bore a suspicious resemblance to the caricature in which the great scientist had been depicted as half-man, half-monkey. Though cleverly done, the trick was eventually revealed for what it was—precisely what many scientists suspected from the off: the cranium of an anatomically modern human (a medieval man, to be precise), placed alongside the jawbone of an orangutan. It was the greatest hoax of twentieth-century science.

In the wake of the scandal, palaeoanthropologist Louis Leakey was convinced that he would find a hominid fossil in Africa. Although from an English family, he was born in Kenya and grew up around Kikuyu, which must explain his instinct to look beyond white European man. The major breakthrough of his professional life, one that definitively turned scientific attention towards East Africa, came in 1959, the centenary of Darwin's *On the Origin of Species*. Though the bones that made Leakey world-famous and palaeontology fashionable were in fact found not by him but by his wife Mary, one day while he was lying sick with malaria in camp. Accounts differ as to his response: did he leap out of his sickbed and race instantly to the dig, or was he in fact disappointed to learn that it was a sturdy *Australopithecus*—which still had not been recognized as humanity's ancestor—and not a hominid? What is for certain is that Leakey was responsible for introducing and naming *Zinjanthropus boisei* (later called *Australopithecus boisei* by scientists and popularly known as "Nutcracker" because of its prominent jaw), and that he did in time begin to value the find, referring to it as his "dear boy".

The Leakey family went on fossil hunting in the Olduvai Gorge through the middle of the century. During that time a theory was gaining currency that our ancestors had become bipedal because of climate change, driven onto the savannah by drought in particular. Palaeoanthropology's golden era in Africa soon followed, with finds in places such as Hadar, Olduvai, Omo and Koobi Fora, many of them in the dried-out beds of ancient lakes and rivers. And there was another difference, particularly marked in the finds in Kenya, Ethiopia and Tanzania: unlike those in Europe and Asia, the African ones were each geographically very concentrated.

As we have seen, the oldest hominid fossils to have appeared at that time had been in the middle and lower reaches of the Awash River, in eastern Ethiopia. But we have not discussed the reason why. This part of the world forms a "Y" known as the Afar Triangle, the result of the triple intersection of the Red Sea, the Gulf of Aden and the East African Rift.

To understand why those fossils appeared there we have to wind the clock back another fifty million years, which was when the drying-cooling of the Cenozoic began, reaching its peak with a series of ice ages around 2.6 million years ago, when falling temperatures reduced evaporation from oceans and rainfall and brought water together in a solid state. It is in this context that the *homo* genus arose. But one thing at a time…

Between twenty-three and thirty million years ago, the tectonic movements of the Miocene period formed the Pyrenees, the Alps, the Himalayas and some of the mountain ranges of South and North America, which were then still separate landmasses. It was a time of temperate climate, but the world then began to cool. Ever since, the world's cold spells have been accompanied by droughts. With the increasing cold and aridity, forests became sparser and woodlands turned to shrubby savannahs. Hence the Miocene is known as the "age of grasses". Around fifteen million years ago the world cooled once more, while continuing to dry out. Some apes developed morphological and metabolic changes that allowed them to adapt to these conditions, leading to the emergence of a new branch: the hominids.

The collision of India and Eurasia that gave rise to the Himalayas was key to the aridification of the world that has been with us ever since. When this mountain chain then began to erode, it lowered global temperatures and reduced the evaporation of water from the oceans, while the interaction between the Himalayas and the Tibetan plateau created the monsoons of India and Southeast Asia while reducing rainfall in East Africa. Meanwhile, a cooler Indian Ocean made for decreased rainfall in East Africa. In addition to these changes, which had global repercussions, others took place in what is now the Afar Triangle, firstly tectonic and then climatic and ecological. Divergence between the African and Arabian plates gave rise to the Red Sea and the Gulf of Aden; the climate, landscape and ecosystem of East Africa changed, and later the forests began turning into savannahs. At the height of a climate crisis around seven million years ago, we separated for good from the chimpanzees with the emergence of *Ardipithecus ramidus*, the hominid ancestor of Ardi, which lived around 4.5 million years ago in the middle reaches of the Awash and is our oldest known ancestor.

Progressive global cooling increased—accompanied by droughts—and at the same time East Africa reached a peak of aridity. The hominin population increased and then ran out of space as the forests receded, driving them onto the savannah. There, at the end of the Miocene, Ardi's descendants—Lucy's ancestors—had to find a way to survive. The new spurs of the Rift Mountains became natural ramparts, sheltering them from wet Atlantic winds. Being in the lee of these great peaks meant far less rain in East Africa, which grew drier around three to four million years ago, while rainfall concentrated further and further offshore. This in turn caused the forests to recede further, giving way once again to savannahs. Our ancestors, who had increased their population before food became scarce, would now have started to spend most of their time walking upright. In addition to bipedal locomotion, they adopted other changes that made life easier for them in this context, as we shall see later on. *Australopithecus* was by this time roaming the world, along with some hominin species we have still to discuss here. Four million years ago, *Australopithecus* split into a bipedal-arboreal branch (*afarensis*) and an exclusively bipedal branch (*anamensis*) which lived in Hadar. Therefore, although Lucy is still deemed the grandmother of mankind for all that she represents, properly considered she is more like our great-aunt.

Our parentage aside, the discovery of Lucy, almost half a century after the Taung Child appeared, meant *Australopithecus* now had to be accepted as more than an ape: what if they were pre-humans, or very early humans? Lucy was just the beginning. On the next field trip to Hadar, Johanson found an entire family. Soon afterwards, Mary Leakey's team came across the Laetoli Footprints, tracks that proved that our and Lucy's ancestors were already walking upright in Africa around 3.5 million years ago. The footprints had been covered in volcanic ash, but the people who left them do not appear to have been fleeing an eruption. It was deduced from the footprints, which are the oldest evidence of bipedalism, that the individuals in question were already accustomed to such a posture and were walking, possibly in search of water, in an unhurried fashion. Other footprints were found next to theirs and initially thought to be those of a bear, though ongoing study of the Laetoli Footprints has revealed that they in fact belonged to another hominin. A species that we have yet to encounter and that was already fully bipedal alongside *Australopithecus*.

Lucy and her kind had swapped treetops for the ground. While some of their forebears, like the Taung Child, had made brief stabs at bipedalism, Lucy spent the majority of her time on two feet, although her brawny arms suggest she had not entirely renounced hanging from the occasional branch. And she might have gone up to the treetops once in a while to sleep, safe in a nest of leaves. But in a place where trees, fruit and water were growing scarce, she would go on increasingly long treks in search of food.

But what is the reason her forebears definitively shifted to bipedalism? An array of theories have considered the fact that it freed up the hands, the energy-saving consequences and the greater ease of communication, and even, as we've touched on, a thirst for blood, but the most accepted idea has to do with the environment I've begun describing, and the demands of thirst and hunger. Lucy and her kind had begun venturing into areas where it rained less, trees were fewer and farther between, and, as a consequence, so were edible fruits and shoots. Extremes of aridity and cold, in other words, seem likely to have driven the shift. What initially did not seem to be an advantage ultimately left their hands free for things like tool-making, as well as reducing their bodily exposure to the sun, while increasing their field of vision on a flat plane with sparse vegetation. But both Johanson and Lovejoy thought they likely first attempted walking before such conditions arose, while they still principally inhabited the trees, before population pressure and the receding forests consigned our ancestors to the savannah. We now know that Lucy's locomotion was almost identical to ours and there would be no major differences between our footprints in the mud and the Laetoli Footprints, which goes some way to proving that bipedal locomotion must have begun much earlier on. More recent research has operated broadly within these parameters; while bipedalism was a key innovation for survival when thirst arrived, our ancestors were already capable of it; they had begun to walk while living a forest existence. And sound forward planning it turned out to be; when thirst came, those who could walk on two feet possibly found it easiest to adapt.

There have been other theories, such as that our ancestors began to walk on their feet because of sex. On the face of it, this is much more appealing than the idea of the killer monkey who stood up to kill. What could drive the evolution of a species more than the reproductive drive itself, if that is what keeps it alive? According to this same theory, without being precisely causal, a series of circumstances led males to set off walking in search of food while the females waited for them to return. The freeing of the hands would have facilitated the exchange of food and—one thing leading to another—monogamy and falling in love would have arisen. It sounds good until one realizes that this thesis attributes the origin of bipedalism to males, relegates females to the kitchen before such a space even existed, and turns the dawn of monogamous love into a kind of prehistoric prostitution in which females offered sex in exchange for food. This idea rather reflects modern-day bias carried back to ancient times.

Australopithecus began to range further afield in search of food, but with less success. How to accumulate energy while at the same time conserving it? Being bipedal brought certain advantages, such as a reduced energy expenditure when moving from place to place, the ability to detect enemies at a greater distance and to carry food. It also made face-to-face sex possible. Who knows if love did not actually come about through copulation with an addition as striking as being able to look your partner in the eye?

But bipedalism was no panacea. Walking on two feet, as Juan Luis Arsuaga put it to the Spanish novelist Juan José Millás in their co-authored books, consists more or less of constantly falling over and catching oneself at the last moment. This adaptation left our ancestors with back and neck pain, with a wider pelvis that helped them stay on their feet but also entailed a narrower birth canal, while the babies they produced had wider and wider heads. Lucy had no such issue: her offspring's brain was very similar to that of a chimp. It was many generations before that organ began to get any bigger.

At some point, being increasingly exposed to predators on their foraging expeditions, *Australopithecus* had to start running. This is where we can relate pork to speed (though with reservations, given how burly and probably slow *Australopithecus* was), as it was around the same time that they began to adapt to food and water scarcity in a similar way to camels. Somehow their bodies "understood" that if they accumulated more fat, they would survive longer in conditions of food and water scarcity.

Lucy was little over a metre in height and barely more than twenty-seven kilos. There is no doubt that she would have known hunger at some point. She had different teeth to those of her ancestors, and a different diet. Though predominantly a herbivore, she could also eat termites, insects and eggs. Her body transformed the fruits she ate into lipids, which she stored in order to reserve energy for times of scarcity. But the storage capacity of *Australopithecus* gradually increased in conjunction with food shortages, when they were forced to survive on roots and to move farther and farther afield. They had also lost the outsized canines of their ancestors, an adaptation that has been associated with the aridity of the environment; in times of drought, their jaws had greater mobility and strength, allowing them to masticate drier foodstuffs.

It is highly likely that Lucy made use of some simple tool of her own making, because we know that the australopithecines before her were already doing so. Instead of keeping and re-using these, as early humans would subsequently do, she would have discarded them along the way. A transition takes place with this tool-making ability that explains why our more recent ancestors did not rely on evolutionary adaptations as much as camels did. But, although Lucy never got to eat eggs and bacon like Don Quixote, without those metabolic, morphological and cognitive adaptations and innovations, today's Manchegans might not have got to do so either.

Lucy was alive three million years ago, when human consciousness was on the verge of appearing. She died, we think, at about the age of twenty-three, and while pregnant. In Ethiopia they have another name for her: "Dinkinesh", which is Amharic for "You are marvellous." The causes of her death are unclear, but those who found her did not rule out the possibility that she had drowned in that place, which could just as well have been a lake or river, and that therefore the water kept her from possible scavengers. That is why her body appeared intact at the bottom of the ravine, when her discoverers passed by millions of years later. It has also been claimed that she died when she fell from a tree, a hypothesis that held for a time, but the palaeontologists and anatomists who found and studied her were strongly opposed. Johanson continued to argue that she drowned in a lake, and that water followed by sand preserved her body until he and his colleagues came along.

Johanson returned to Hadar in 1980 after a three-year absence. In the place where he had found Lucy, there had been so little precipitation in the intervening period that his own footprints were still visible in the sand.

2

Homo sitibundus: the great journey

Everything that characterises us, standing upright, the omnivorous diet, the development of the brain, the invention of new tools, would all result from an adaptation to a drier environment. […] You will say that I am exaggerating, but love is also the result of that drought.

YVES COPPENS,
THE MOST BEAUTIFUL STORY IN THE WORLD

We believe that North Africa was a very dry place when the first humans began to leave it behind and spread around the world. It was the conditions in a green Sahara turning more extreme that prompted our ancestors to leave the continent.

INTERVIEW WITH JESSICA TIERNEY IN
NATIONAL GEOGRAPHIC

It was a winter's day. The men were beating the branches of a Manchegan olive tree, while the women knelt to pick up any olives not caught by the shawl spread out on the ground. One of the men, looking for an excuse to strike up a conversation with one of the women, got down on his knees as well and started to gather olives. Another of the women, interpreting his move and wanting to speed things up, said to her sister, "It's you he's here for." Under the olive tree, the pair began to talk. At the disco he wrote her a note suggesting a first date by a water catchment tank, in the style of those ancestors who flirted beside wells in the middle of the desert a million years ago. In arid environments, courting by the water's edge must itself constitute a kind of promise.

Three years after that first encounter, she gave birth to a girl in a nearby village. It was a place surrounded by vineyards, which had sprung up there for the same reason as the olive trees. The country people in the epicentre of dry Iberia, beset by water scarcity, found in olives, grapes and cereals a way to say to the land that they would not be moved on, and neither would their children or grandchildren. Ever since then, olive trees have provided a link between the living and the dead. In my case, the olive tree has other implications besides, and I sometimes wonder whether my brother and I would have come into being had those young pickers—our parents-to-be—not met on that winter's day.

The furthest back I can reach in my family tree is a bare ten generations. The oldest baptismal and death certificates in the church archives go as far as the mid-sixteenth century, beyond which lies a yawning void of thousands and thousands of years. Myriad anonymous women and men. But there are invisible texts older than the written word. I don't know how to read them, but the people conversant in their language have taken it upon themselves to translate some fragments. Here I mean geneticists; thanks to analyses of ancient DNA and their efforts to decipher them, I'm able to fill in the very first (or last) gap in my maternal line. In the year I came out of my mother's body, a group of researchers who had gathered samples from around the world and sequenced their mitochondrial DNA—which is passed down the maternal line only—identified the mother of all living people on Earth. With the information they obtained, they created a kind of maternal family tree and located the mitochondrial "Eve" somewhere in Africa.

It was not scientists, but journalists who gave her a biblical name, a somewhat misleading one: mitochondrial Eve was neither the only nor the first woman to have children; at that time and even earlier, there had been others both in that locality and elsewhere in Africa. But the mitochondria in her ovules were the only ones that endured, while other lines became extinct or ceased producing daughters. For the story to be completed, predictably enough, the search for Adam, with his paternally inherited Y-chromosome, then began. But there was neither baobab tree nor well to bring them together, nor snake to tempt them, because the male ancestor lived in another part of Africa, and it would have been a strange miracle if they had coincided, since in principle a span of 50,000 years divided them. In a seeming attempt to bring them together, new research has shortened the time between one life and the other with the same success as those trying to prove that Leonardo di Caprio's Jack and Kate Winslet's Rose could fit on the same raft in *The Titanic*. It doesn't seem possible for them to have coexisted.

Subsequent studies looked for a place on the map for Eve and attempted to adjust the dates. The conclusion here was that she lived around 150,000 years ago in a part of southern Africa bordering northern Botswana, in the Zambezi River basin. Long ago, a trio of rivers (the Zambezi, Okavango and Kwando) became landlocked, giving rise to a lake the size of Switzerland—the largest in Africa. But the Makgadikgadi later began to recede and, as it broke up into several smaller lakes in turn, created a fertile marshland that supported the anatomically modern humans who lived there for 70,000 years, until changes in climate drove most of them elsewhere. Eventually the water evaporated and salt flats appeared. When mitochondrial Eve lived there, the lake surely had a different name: Makgadikgadi is Tswana for "even drier dry place". Today it is a white spot in the middle of the Kalahari Desert. This is what a team of geneticists, geologists and climate physicists discovered, but their findings were widely criticized; other scientists see mitochondrial DNA as insufficient to establish the origin of humanity. But the researchers in question replied that this wasn't their aim; they were merely trying to understand what life was like in one of the homes—not the cradle—of the first anatomically modern humans.

They concluded that some of mitochondrial Eve's offspring would have migrated in several waves as the rains returned and green corridors opened up. The first wave went northeast about 30,000 years ago, while the second went southeast about 20,000 years ago. But not all of them left. Some of their descendants stayed on as hunter-gatherers, and are there to this day, in a place that eventually became a salt desert. People with whom many of us in Spain share DNA—predominantly the Joisan, the result of the union of the Joi and San ethnic groups—are present all across Africa. In the Kalahari today, where they are most concentrated, barely 50,000 Joisan remain, although it is thought that the origin of the L0 lineage may be in East Africa. Their search for water, which sometimes turns to desperation, inequality and injustice, has continued to corral those who stayed behind. The San, having been driven from their lands, took on the Botswana state and recently won the right to return to their native homes. In reality, however, not all the exiles were granted this right, but only those who had engaged in the legal battle. The few allowed to return were faced with two constraints on their way of life and survival: they were banned from hunting and from accessing their own waterholes, while resorts built swimming pools for tourists—who were also given hunting permits. Although they eventually managed to regain access to water, their way of life is on the verge of disappearing.

In spite of the immense void in my family tree, and although I cannot link it to all the women of my lineage who have gone in search of water, I can connect the first to the last. To me it is endlessly curious and terrifying that my closest and most distant ancestors, separated by more than 150,000 years, 12,000 kilometres and around 6,000 generations of women, were both born and lived in places whose toponyms signify basically the same thing. Despite the distance, La Mancha and the Kalahari share semantic origins: one is "dry land", the other "place without water". The Kalahari has even been given other, more poetic meanings, such as "great thirst" and "to be in a place so dry and inhospitable that one becomes thirsty". For that is what became of mitochondrial Eve's supposed Garden of Eden, in which chromosomal Adam may never even have stepped foot.

The search for water even gives clues to the origin of the genus *Homo* and of the species *sapiens*. Some time after the rift opened up on African soil, thrusting new mountains into existence and reducing rainfall down the entire east side of the continent, the peoples who remained had to find ways to survive extreme conditions, and in their wake came *Australopithecus*, Lucy included, and, some time later, Eve's ancestors.

This was the view of palaeontologist Yves Coppens, who proposed that bipedal hominids (what we today call hominins) originated in East Africa. He called his hypothesis *East Side Story*—it being a common strategy among popularizers of science at that time to make puns on the titles of musicals, films and the like. The basis for his argument was that the oldest ever fossils had been found in the eastern part of the Rift Valley, which is an arid zone of savannahs and open woodland. The fossil record proved him right for a time, and the scientific community, which had been debating between an African and a multi-regional origin, began to accept it as the birthplace of humankind and of our entire species alike. The earliest remains of the genus *Homo* as well as of the species *sapiens*, and of their ancestors, were found there. That is why Coppens located his "(H)Omo event" on the Omo River. At least, he did then, and for a time the view held.

Coppens saw Lucy as sometimes walking on the ground and sometimes climbing trees, because her life was spent between forest and steppeland. "Then the earth tilts back on its axis and drought liquidates forests and fruits," he writes. It is then that Lucy, her people and those who come after them, their steps short and tottering, have no choice but to walk upright almost all the time. A couple of millennia after Lucy's death, and with the emergence of the *Homo* genus, the Pleistocene came in. The world entered a cyclical cooling-warming pattern. As ice accumulated and partially dried up the oceans, aridity increased. Then came an interglacial epoch, followed by a new ice age. Another devastating drought around 2.6 million years ago wiped out many species of animals and trees, while fostering the growth of grasses. It was then that humankind emerged. This is important in understanding all that followed; after all, if adaptable grasses such as wheat had not flourished in arid areas, there would be no such thing as beer, pizza, sushi or tacos. Modern-day consumers live off products made from grasses as well as the meat, eggs and dairy produced by animals that feed on grasses. Grasses and ungulates—hoofed mammals—appeared in the world around the same time.

So it was then, as aridity pushed *Australopithecus* to forage for roots and hard fruits, that early humans learned the advantages of flexibility and adopted an omnivorous diet. They developed more elaborate tools and would have discovered new emotions and feelings. In terrain where aridity laid waste to much vegetation, babies were more exposed. Their vulnerability brought mothers and fathers closer together. Might it have been that love now also emerged? All this, Coppens said, "would result from an adaptation to a drier environment", but what came was the ability to swiftly adapt to *any* environment.

Evolution is more alike to a tree than an arrow: one species need not necessarily die out altogether to make way for a successor. There are three million years between the grandmother of humankind and the first *Homo sapiens*, which does not mean that *Australopithecus* did not coexist with early humans for a time or that early *Homo sapiens* never coexisted with other human species.

Our African ancestors inhabited a vast territory abounding with rivers and trees, in which for example the Sahara as we know it had not yet formed; it was pre-dated by a well-stocked fruit garden, its rivers teeming with fish. Changes in climate both drove those ancestors apart and brought them closer together. The dance between desert and forest commenced. The former began to spread, swallowing up the rivers and increasingly displacing humans.

Several more primitive groups inhabited Africa alongside *Homo sapiens*. They already had tools and implements that allowed them to hunt in a more sophisticated way, which they did alongside gathering and possibly also fishing. They led a nomadic life, moving from place to place in search of food and water. Thanks to the legacy of their ancestors, they could walk on two feet without appearing drunk. They no longer struggled to keep from falling over as one foot travelled through the air and landed on the ground again, and their digestive systems allowed them to conserve energy. All this in an environment dominated by cold and aridity. They moved constantly because the cold and drought left them with only three options: adapt, leave or die. The former had already happened. Although the debate is ongoing, several authors see cold and drought as the origin not only of bipedalism but of important later increases in brain size, known as "encephalization". Before long, the brain of *Homo erectus* was considerably larger than that of *Australopithecus*.

How did the rest of the body react to this development? The view amongst scientists tends to be that one organ becoming so much larger, with the energy expenditure that entailed, must be detrimental to all the other organs. In the late nineteenth century, Sir Arthur Keith believed that the brain had grown at the expense of the gut, which inevitably shrank to compensate. Fast forward 100 years, and Aiello and Wheeler concluded that brain growth would have been impossible without a change in diet. Fewer plant foods, more animal-based foods. Two changes ran alongside that one: the scarcity that arose due to the climate, and latterly the newfound capacity to cook. Other modifications followed: the larynx moved lower down the neck, which in turn allowed vocal cords to take shape. And with the necessary cerebral connections in place, one of our greatest tools arose: speech. Along with brain development, here now was the basis for complex language. The first words would have been spoken. With more complex tools to wield, hunting itself became more complex. Clothes had to be sewn, and the use of fire had to be perfected, and meat cooked. And along with fire, surely, came storytelling.

But to zoom out a little: cognitive developments, driven by factors such as thirst, cold, hunger and an omnivorous diet, enabled small pointed tools like awls to be made and in turn needles for sewing garments. Water vessels and objects for things like cooking and burial rites soon followed. Things that had existed previously only in the world of imagination and story took physical shape.

Where did all of this take place? Coppens's ideas were for a long time upheld by the evidence. But he himself went on to find *Homo ergaster* fossils in central Africa and *Australopithecus* fossils in the wetter parts of the Rift. Fossils of *Australopithecus bahrelghazali* appeared on the far side of the valley, in Chad. Further, although the oldest *Homo sapiens* remains were found in the Kibish cave (Ethiopia), earlier equivalents, including fossils dating back 315,000 years, have been unearthed as far away as the Jebel Irhoud site in Morocco. Although they were first identified as *Homo neanderthalensis*, they were later considered *Homo sapiens*, which then brings forward our species' origin date.

The fossil record was sparse and scattered, hence several hypotheses continued to arise after it was widely accepted that we came from Africa. Palaeoanthropology needed something to supplement it, and population genetics provided that something. This discipline has the advantage of not depending on the hypothetical appearance of physical remains to expand on and contrast information, based as it is on something immaterial that has not been broken down by time: our genes. These, in turn, have been combined with fossils to reinforce or refute certain hypotheses. With the discovery of mitochondrial Eve, for example, and despite the hypothesis being highly controversial, the study of genetics was combined with archaeology and palaeontology to deepen our knowledge of humanity's origin, and this has only reinforced the idea of Africa as its point of origin.

While palaeoanthropologists and archaeologists constantly pointed to Ethiopia, Chad or Morocco, genetics pointed to the south of the continent. All these findings led the scientific community to take up the "Out of Africa" hypothesis, which located anatomically modern humans there before they scattered across the globe. In recent decades, various finds, studies and genetic analyses have suggested different places on the continent, but researchers then realized there was no need to place a *single* pin on the map. Since then, various places in Africa have been accepted as origin points, with the idea that our ancestors drifted between them.

It was once accepted that *Homo sapiens* began migrating from Africa when climate conditions allowed for it. That is, when the desert receded and corridors opened up. But now that archaeology, palaeoanthropology and genetics have been coupled with geology and palaeoclimatology, more recent findings suggest the opposite: our ancestors would have started to leave, and continued to do so, at times of prolonged drought, which created deserts where once there had been lakes. This is also compatible with the so-called *hydro refugia* hypothesis, according to which the dispersion of anatomically modern humans could have been oriented around springs. Whether they waited for climatic conditions to improve or left during drought, it seems clear that their horizons expanded at times when the climate was changing. If they were not thirsty at that precise moment, they certainly had been at some point.

One day, about 120,000 years ago, two or three individuals made their way to Lake Alathar, a place frequented by thirsty elephants and camels. For some unknown reason, the humans turned back. But by now that lake was in any case a semi dried-up mudflat. When it receded fully, footprints were left forever in what is now the

Nefud Desert in Arabia, bearing witness to a long, long-ago walk out of Africa.

For a footprint to fossilize and survive to the present day, water must first be present and then disappear for a long time. This tends to happen in deserts, but ones that have not always been dry. Such traces, called "ichnites", are thought to be a response to increasing aridity. The individuals in question would have arrived at a time when the desert was turning green once more, and therefore drawing in thirsty humans and elephants alike.

At that time there were two ways to leave Africa, apparently the same routes as those taken by *Homo erectus*, who made it to modern-day Georgia almost two million years ago. In warmer, rainier periods, the northern route—eastward from the Sinai Peninsula—was an option. But in times of drought, the best was the southern route, which passed through the Bab el-Mandeb Strait on the Red Sea and led to the Arabian Peninsula. During the ice ages, drought made the desert exit unviable, while the lowering of the waters made passage through the strait easier. And then in interglacial periods the latter was blocked by rising waters, while at the same time the desert regreened, making possible a way out along the banks of the Nile.

The Arabian footprints do not denote our species' oldest-known attempt to leave Africa; fossil traces found in modern-day Israel and Greece predate them by 200,000 years. Various attempts were made around that time, coinciding with population increases that in turn reduced the amount of available fertile land for settlement. But rarely were these a success; short in duration and distance, there are only token instances of them in the fossil record, and none at all among the current-day genes of peoples born beyond African shores. Whether those attempts failed because the individuals involved died, or their offspring died, or they turned back somewhere along the way, we do not know. And why would they have turned back? Some kind of Ulysses syndrome seems less likely than the possibility of meeting and being repelled by burly *Homo neanderthalensis*, who had already made Eurasia their own. These comings and goings may have mirrored their advances and retreats on what was by now their home continent.

A short time after the journey to Arabia (well, 30,000 years later), a descendant of mitochondrial Eve began to walk on two feet and left Africa. One small step for her, one giant leap for humankind. With a handful of others, *Homo sapiens* finally embarked on the first stage of their long journey to the rest of the world. It is thought that only about 1,000 anatomically modern humans set out, moving slowly as they fled drought and cold, while following the movements of the animals on which they fed. Some would have followed the route of their *Homo erectus* ancestors, perhaps using sledges. Others took advantage of the temperate moments when the desert was greening and went the northern route. After several attempts, some 100,000 years ago another wave finally opened the door to Eurasia. This time they did not look back. Some of those who had left Africa via the Sinai Peninsula ended up in the Near East. Others kept on going, passing through Arabia and stopping in Anatolia, then a sort of bridge between present-day Turkey and Europe. In the Near East, some of those disparate groups may have met once more, only for their onward paths to diverge. In present-day Iran and Iraq, they branched off into Southeast Asia and Europe, the latter still inhabited by *Homo neanderthalensis*, although they were by now in decline. By about 45,000 years ago, some of the wandering *Homo sapiens* had made it as far as the Iberian peninsula.

Those who had gone to Asia went on to the Indochinese peninsula, and from there to New Guinea and Australia. Those who had

HOMO SITIBUNDUS: THE GREAT JOURNEY

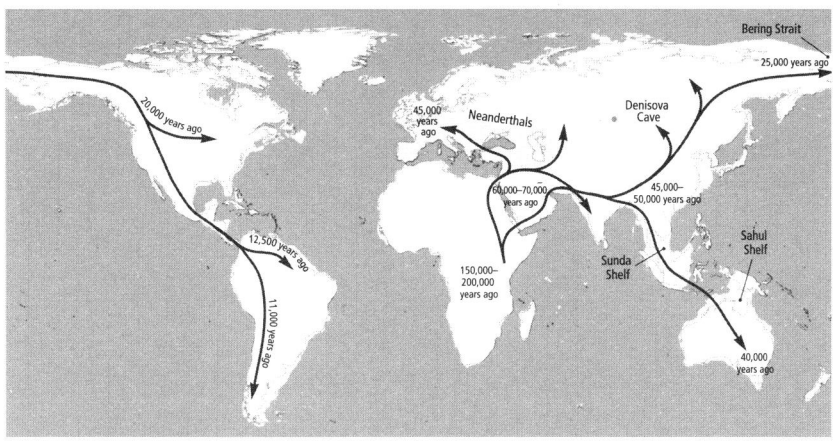

Migration routes of *Homo sapiens* during the Ice Age and approximate distribution of *Homo neanderthalensis* and *Homo denisova*.

turned north, with the help of the ice—which dried up the seas and caused the land to rise—crossed the Bering Strait and reached North America and, finally, South America. This latter group, unbeknownst to them, had made the reverse journey to that of the camels which left the Americas via the Bering Strait long before, and ended up in Africa.

Although the date of *Homo sapiens*' arrival in South America has recently been brought forward, modern humans had by that time already spread to almost all parts of the globe except Antarctica, New Zealand and certain islands. Surely our ancestors were not looking for territories to conquer and explore: their only aspiration was to stay alive, and they moved slowly, in accordance with droughts, rainfall and the migrations of the animals they hunted. Having settled in new and fertile lands, they began to multiply and found themselves once again in the same situation that had driven them from Africa: with not enough fertile land or fauna for all. This was the beginning of a new expansion which, with a few exceptions that still exist today, was to touch them almost universally: that of agriculture.

Our first great journey is only a "journey" if we put the whole thing on fast-forward. The reality was very different, far slower. Like the tree that, without anyone noticing any movement, ultimately does move the forest by scattering its seeds, anatomically modern humans reached every corner of the world. They made the journey not as individuals, but as a species, since in reality each generation would have moved only a few kilometres.

Although there are no fixed dates for the stages of this journey, because each new find brings with it new possibilities, it is now estimated that over the course of about 80,000 years our ancestors had spread throughout almost the entire world; also that the point of departure was not a single one, as was once thought, but multiple points in Africa. It is not that science lies, contradicts itself or is consistently wrong: it is rather a constant dialogue in which no one ever completely stops talking, and of which we are simply the fortunate witnesses. It is possible that, between the writing of this book and the time you hold it in your hands, a new fossil will come to light, or a DNA or ancient pollen analysis, and with these all our certainties will go out the window. Back to square one…

In the meantime, what we do know is that modern humans began to appear in Eurasia less than 100,000 years ago and that, after tens of thousands of years of slow dispersal, they were greeted expectantly by *Homo neanderthalensis* and *Homo denisova*, and that, whatever the nature of the contact, *Homo sapiens* became intimate with both. Whether the encounter involved romance or violent misunderstandings we do not know, but it is none the less engraved in the genes of people all over the world today. Both sets of hosts were, in fact, distant cousins: descendants of the ancestors who had left Africa almost two million years ago either by those very same routes or, shortly afterwards, via the Strait of Gibraltar.

From a corner somewhere, from one cave or another, *Homo neanderthalensis* looked on dumbstruck as strangers started showing up unannounced and with no apparent intention of leaving. These were more lithe and darker in complexion. When *Homo sapiens* entered Europe, they found it populated by short, muscular, thick-set individuals with broad noses, forearms shorter than their own and lighter skin. *Homo neanderthalensis* had been there for some 200,000 years. During the most recent ice age, in a world of extremes, the Iberian peninsula became one of the few habitable places for both. While *Homo sapiens* weathered the terrible conditions, the last *Homo neanderthalensis* appear to have perished in this same corner of Europe.

We once thought that *Homo neanderthalensis* became wholly extinct and that this might have been the doing of our species. They were already in a period of decline, however, their population having dwindled, and were also spread out in isolated groups plagued by inbreeding. Besides, would one species who set out to systematically exterminate another live alongside it for thousands of years? Considering what we see in today's world it isn't so hard to believe, but if we put ourselves in those people's place—aside from feeling bitterly cold—we would probably find the *Homo neanderthalensis* fairly frightening to look at, so muscle-bound were they and simply different to anything we had ever seen before. Although they were strong, they lived short lives and might not have been as flexible as *Homo sapiens* when it came to adapting to changes in climate; the latter had fitted themselves to a range of different environments by now, partly by being flexible in diet, which allowed them to feed themselves wherever they went.

In recent years, interest in *Homo neanderthalensis* has grown exponentially. Perhaps the near-daily articles about them continue to appear because we now know they never became fully extinct: the genes of *Homo neanderthalensis* live on in present-day humans, most of all in Eurasia. Doubtless headlines in the western media will soon reflect new knowledge about *Homo denisova* genes—which enable people to live at high altitudes—being present far beyond Asia.

The idea that Neanderthals and Sapiens interbred was explored in the 1911 Belgian novel *The Quest for Fire* by J.-H. Rosny. The film adaptation caused a stir because such a thing seemed unthinkable, and to some almost an insult. Even Coppens, who defended the intellectual, creative and emotional capacities of Neanderthals, considered the idea of their having mixed with Sapiens nonsense. But science has gone on to prove the fiction correct: geneticist David Reich contends that this mixing is limited only due to the later migratory flows of Anatolian farmers, whose expansion diluted the Neanderthal inheritance. Interbreeding was reduced for social reasons, contrary to other researchers' suggestions, namely low fertility among the hybrid offspring. We may not even be different species, given that they interbred and that some of their descendants managed to continue procreating. In fact, some researchers prefer *Homo sapiens neanderthalensis* as a classification.

We now know that Neanderthals were not as callous and violent as was once thought. These seemingly coarse, strange-sounding natives, often depicted as akin to trolls, also cared for their elderly and sick, buried their dead and were conceited in their own way, in that they had an aesthetic sense that meant they collected shells for pleasure, made necklaces and painted on cave walls. They made primitive musical instruments, as well as the oldest-known kind of bread, and were gourmets who held seafood feasts in caves overlooking the sea (no one bothered to clean up afterwards). They spoke, in their way, though it was a way we might not now understand. They may even have fallen in love. At some point, the newcomers discovered that they had more in common than not.

Some have suggested that *Homo neanderthalensis* left us an inheritance of artistic sensitivity. Their brain was larger than ours are today—they had no choice but for it to grow: the most recent generations, without brains that allowed them to turn pelts into warm garments and caves into decent shelters, as well as to keep fires going both for warmth and cooking, had only just survived the cold. Their physical appearance had been shaped to some extent by the climate: their huge nostrils and flared cheekbones meant that the air took longer to travel down into their lungs, and would have been warmed and moistened along the way. This bone structure meant that they could withstand the freezing, dry air without difficulty breathing. Their small stature and stoutness, meanwhile, combined to help them retain body heat. However, not everything they passed down to us was a blessing, for Neanderthal genes also expose us to a tendency towards depression, arthritis, allergies and a weakness for cigarettes. While they have endowed us with immunity to some diseases, severe cases of COVID-19 can be worsened if one has a Neanderthal inheritance.

The history of our species, as well as what we have inherited from others, could be summarized as a series of responses to changes in climate, and especially to droughts. Some of these responses seem absurd at first: standing up or working in the fields may not have seemed like the best options initially, but over time they became so: both allowed us to sustain ourselves in arid places. From leaving Africa in search of fertile land to the construction of reservoirs, so crucial in modern agriculture and industry, stopping off along the way with the invention of agriculture, the shift from nomadism to sedentary living and the building of the first cities, we have done nothing but adapt to climate and aridity—or indeed force nature to adapt to *us*. Where once astronomical changes altered the Earth's axis of rotation, which in turn altered the climate, today it is thirst that is having this effect. Today's thirst may be the cause of tomorrow's. There will be those who will call it "drought" as a way of avoiding human responsibility—disavowing environmental determinism while in fact falling into it.

3

Water lessons

> Water, is taught by thirst.
>
> EMILY DICKINSON

In summer and early autumn a strange scene takes place in certain arid areas. At daybreak, without fail, a male Iberian sandgrouse sets off to the nearest waterhole, sometimes a flight of several miles. As it flies, it calls to rally the other males, who duly fall in. One of them acts as outrider, checking for predators and, once the coast is confirmed clear, he calls out to his companions to join him at the water. They all land and begin to bob up and down on the water. Aware that their crop feathers are more sponge-like than the rest, while they bob away, they scrub at that part of themselves, to retain as much water as possible. Feathers duly saturated, they fly back to pass the water on to their chicks.

The journey there was a honking tumult. The one back is silent.

On his return, the male does not go straight to the empty burrow or tree hollow that serves as the nest, but stands some distance away, sometimes on a rock, again in case of lurking predators. He stays there, and the female herds the chicks over to him and his sopping feathers. When they get to him, the male remains upright, thrusting out his crop, and the chicks flock to him in search of water.

Only in the cool of morning does this happen. After that they all hunker down, trying to survive the soaring temperatures. Then open their beaks and emit a guttural panting to cool themselves down. Sandgrouse live with water scarcity in the semi-deserts of Africa, Madagascar, Asia and across Europe. The Iberian sandgrouse, with its tough, cheap meat, has given the whole Spanish-speaking world the expression "it's a sandgrouse"; that is, a giveaway or a bargain. They have found ways of surviving in drought conditions by adapting their routines. Water scarcity is so common in these places that they have learned to be guided by it; they become disorientated when the water holes are fuller than usual. They make the trip at more or less the same time every morning, without any suggestion from the chicks that they are in need of liquids. The question this brings us to is: how do we first begin to ask for water?

Our search for water, as we have seen, began after we emerged from the water, a shift that was integral in the story of life on Earth. But its story, the one we can tell, started when a baby opened its mouth and spoke its first word. It doesn't matter which baby. Or where, or when. Only that the baby in question will have been adorably tiny, with eyes disproportionately large in relation to the rest of its body—proportions that are nothing if not calculated, and common to other mammal offspring over millennia in making them more loveable; proportions that protect us from being killed at birth, that cry out to be held close, to be called "beautiful", all eliciting the universal cooing musicality of maternal language. Once it is thirsty enough to speak, the creature finally says, "Water!"

The Dogon people of Mali believe that the first word came about to bring order out of the world's chaos. But what was the word? I have not spent much time around babies, but enough to witness the primordial pattern that drives them to ask for water or cry out for their mothers. Thirst makes seekers of us, it drives us, it asks of us; it asks us to do, to ask, to seek. Thirst is like the dopamine that both urges us on and, once the urge is met, is again secreted. An endless cycle.

Human beings appeared in the world and are believed to have acquired consciousness around 2.5 million years ago. What we do not know is when they first asked themselves the questions about existence we are still asking ourselves today. Sixty thousand years ago is often given as a guide, with the appearance of figurative rock art and all its seeming symbolisms. At that time, and over the following 30,000 years, something changed in their brains and in their lives. A period, as we have just seen, that coincides with humankind's great global dispersal. The most accepted idea is that a genetic mutation paved the way for the brain to grow in size and complexity, allowing for the emergence of a more elaborate spoken language. Some scientists believe that this mutation was in a sense forced to happen, by dint of a cold and arid climate.

Whatever the case may be, it seems that *Homo habilis*, the first of the genus, uttered the first words almost two million years ago, but it wasn't until much later that our ancestors became sufficiently articulate to come up with and share stories. This was no sudden change, however. It has also been suggested that the shift from a simple to a more elaborate language may have been triggered by a need to encourage cooperation between individuals—by means of the gossiping kind of storytelling. At that time another sort of social glue emerged that would not have been possible without such a language: shared myths and beliefs became essential to bring the group together under the same umbrella. Humans had for a long time been endowed with symbolic thinking, and had managed to fashion tools that first appeared to them in their imaginations, but now they were able to relate in words both what they saw and what they did not see, and they also understood the importance of this sharing. If cooperation was key in our evolution, as is widely believed, two major evolutionary steps came with the ability to create the sort of shared stories that formed early beliefs and origin myths, not to mention gossip. All of these could have arisen at night around a campfire. But they could not have come into being without the first word.

What words have in common is that they die. They tend not to last for more than 10,000 years. But someone asked what the oldest one was, and came up with twenty-three. You, I, we, them, no, this, that, who, what, old, black, male, mother, hand, fire, bark, ashes, worm, give, hear, throw, flow, spit. All of these, it seems, have been with us for some 15,000 years. But there is one that may have come about much earlier. Linguists suspect the very ancient spark may have been "no". Perhaps a certain *Homo habilis* telling his son not to do something that might upset people. And is there no trace of water among the oldest words we have preserved? Well, one of them is "flow". Such a hip word nowadays, yet surely that putative first speaker said it while pointing to a trickle of rain or a river, or maybe in surprise at what she suddenly found falling from the sky after a long drought. *Pluvia*, *pleure* and *plovere* in Latin precede *lluvia* and *llover*—Spanish for "the rain" and "to rain"; the Indo-European root is *pleu*, "to flow". *Plou poc, però quan plou, plou prou*, say people from Valencia and Alicante; literally—and far less alliteratively—"It rains little, but however little it rains, it rains enough," a play on words that perhaps speakers of various Indo-European-rooted languages could understand, thanks to the legacy of the Yamnaya, whom we will meet later on. Water must have been important to speakers of proto-Indo-European languages, since what remains of their heritage in our language in relation to the land revolves around it entirely.

Agua—"water"—comes from the Indo-European root *Akwā*. If I think of it in other languages, I hear the babbling of a thirsty baby. *Water. Eau. Aigua. Auga. Apa. Acqua.* Even an Armenian exception, *jur*, brings me to the protolanguage of infants, with its seemingly universal "goo-goo ga-ga" sounds. *Guagua* in Mapudungún, the language of the Mapuche, means "baby" because they believe it to be the first sound we make; in Quechua the same word means "child". In some places in Mexico, *guache* is a nursing child. In my village, *guacho* is a motherless sparrow chick that has fallen from a roof. Gua is also the name of the first chimpanzee on which verbal experiments were carried out.

The lack of clear evidence for how human language first came about meant trying alternative approaches, not all of which had the universal support of scientists. One of these has been to study the way in which other animals communicate, and specifically apes, as such close cousins. In the early twentieth century, researchers began looking into whether non-human primates could communicate with each other as humans did, and with humans too. The experiments they came up with are, to the modern mind, horrifying. With the aim of determining whether language is innate or cultural, various chimpanzee babies were wrenched from their natural environment and sent to live in human homes, to see if they could be educated like human babies. At seven-and-a-half months Gua was taken to live alongside a ten-month-old boy, Donald. Gua learned to use a cup and spoon long before Donald did. When Donald was responding to only three requests from his parents, Gua reacted to more than thirty. The father publicly acknowledged that, after a year of living together, it was clear that Gua was "smarter" than his son. But then Donald started talking and Gua continued to make only the odd sound. Thereafter, however, Donald continued to be able to say just a few words, at the time when children his age had mastered around fifty. They halted the experiment on discovering that their son was imitating the chimpanzee's sounds; Donald had grown to feel that Gua was a sibling. Gua then had great difficulty adapting when he was returned to his natural habitat. The parents were accused of traumatizing both child and chimp, of "making a monkey out of a child, out of a desire to make a child out of a monkey". Gua died within a year of being returned to the wild. Donald committed suicide shortly after his forty-third birthday, months after his parents' death.

A few years later Viki, another young chimpanzee raised with humans, managed to say four words: "mummy", "daddy", "up" and "cup". That was all. From these experiments, language was understood to be beyond chimpanzees. Scientists initially put this down to a lack of intelligence, but soon concluded that it was due rather to an anatomical limitation related to the larynx. Even speech therapists could not get them over the line, and attention turned to other ways of communicating. An American couple, both psychologists, adopted Washoe the Chimp and tried teaching her American Sign Language (ASL). By imitation, Washoe learned the signs for more than 350 words and showed the ability both to create new words from those she already knew and to teach them to other chimpanzees. She even managed to form a number of simple sentences.

When she first saw a swan, she put together two signs: "bird" and "water". After five years in a human home, she was sent to a lab for further experiments, and her full potential was unleashed. She learned a further 250 signs and demonstrated that a chimpanzee is able to state that it is crying in sign language when it receives bad news from humans. That is, she had learned that this is the usual human response in such situations and so set about *explaining* that she was crying, without actually shedding a tear. A show of empathy that must have been a great lesson for those who had forcibly removed her from her environment.

Experiments went on being carried out both in private homes and in the chimpanzees' own habitats, as researchers were determined to prove that complex language could not be the preserve of humans. Another researcher couple spent four decades learning the language of these apes, building on the work of those who had attended Washoe's human language classes. They concluded, fascinatingly, that apes have their own syntax, are capable of lying and, given their ability to alliterate in sign language, also of composing poetry. Today there is even a dictionary that lists the meanings of their signs.

Something similar happens among gorillas. Although they have a peculiar language based on burps and belches, as discovered and mastered by Dian Fossey, it became clear that when communicating with humans they do better with signs. In the 1980s, the world was introduced to Koko the gorilla. When Fossey taught Koko ASL, expecting her to learn only a few words, over the course of four years she mastered close to a thousand signs. The first was "to drink". When conversing with a human who asked if she was going to become a mother, she replied, "Koko-love… sip". When asked about her favourite food, she replied: "to sip". Fossey then asked Koko what her favourite drink was: "drinking apple". She also loved ice cream and formed an amusing way of ordering it: "my cold cup". By the end, Koko could comprehend 2,000 human words. To top it off, she taught around 750 signs to a gorilla she lived with for a time.

Similar experiments were carried out on less famous orangutans, but in 2012 one called Rocky was the subject of a study published in *Nature*. Scientists concluded that the responses he gave to humans were entirely divorced from the language he used with others of his species. It turned out that he was able to create different sounds to respond to humans and that he could also make different "wookies"—as orangutan vocalizations are called—to mimic humans making wookie-like sounds. Subsequent studies have led scientists to believe that the consonants we use come from the arboreal ancestor we share with orangutans, since only orangutans produce consonants, unlike primarily ground-dwelling apes.

So chimpanzees, orangutans and gorillas have been shown to be able to communicate with us. Why, then, do they not talk? Because, it seems, they lack a series of brain connections to the larynx and tongue muscles that enable us to speak. Although scientists have yet to identify precisely where these are located, archaeologist Gordon Childe proposed the idea almost a century ago, and was convinced that the connections in question were just above the ears.

It is curious that one of Gua's, Viki's and Koko's main interests when they arrived in the human world were containers for liquids. Lucy, another chimpanzee, was also fixated by a certain cup, which she called her "red drinking cup". Chimpanzees, like us and other apes, drink water from cupped hands. Pots first appeared about 10,000 years ago in Anatolia and were preceded by eggs. Did the egg or the hand come first? Did the hand evolve to hold the egg that became the pot or cup, or did they fit together like two perfect puzzle pieces?

In some places the constellation Ursa Major is called the Big Dipper, and an Anglo-Saxon myth has grown up around it, telling the story of a girl who sets out from a drought-stricken village with a cup in her hand—hoping to bring water back to her sick mother— and ends up inhabiting the sky realm. Vessels and bowls have both been present in life and death for as long as we have had the capacity for symbol-making. They appear alongside bones in prehistoric burial sites, and still today a cup will be placed next to the deceased in some Romanian villages, just as a bird bath is placed in Muslim cemeteries. Oded Galor says that, in addition to the brain, it is the hand that sets us apart from other mammals, because it evolved "partly in response to technology, specifically for reasons of creating and using tools for hunting and needles and utensils for cooking". So why is it that other apes use their hands as humans do to drink water, and when they manage to communicate with people, why are they quick to ask for a cup?

At one time it was even believed that the lines of the hand functioned like the watersheds of rivers when the hand became wrinkled from wetness. Eventually it was shown that the exact opposite was true: the lines do not carry water inwards, but outwards. This discovery, after observation of people's wet hands, led humans to invent tyres with a special kind of grip that made them safer on rain-slick roads.

What if we first started talking because we were thirsty? According to Rousseau, that would be nonsense: language, in his view, was unnecessary for the fulfilment of most practical needs, but necessary for expressing emotion and for persuasion. "Neither hunger nor thirst, but love, hatred, pity, anger wrested the first voices from [early humans]." I am among those who believe that language came into being gradually and that the exact origin for its current-day form is impossible to find. But the first word must have had something to spark it. We may have started talking for the same reason that babies, cuddly toys and cartoon characters have big eyes. There they are: you, me, we, hand, give… The philosopher Jaime Nubiola says that if a child sees a flower, points to it and says "flower" to its mother, this is in order to bond with her. And this is the essence of the cognitive revolution, which created the need to bond through spoken language. Perhaps, unaware that plants developed flowers as a manipulative strategy, to attract herbivores and spread their seed, the child is only trying to nurture love by sharing something beautiful. But by the time that occurs, it will probably have already said "water". This idea ties in with another of the many theories about the origins of language, according to which mothers could no longer constantly hold their babies while gathering fruit or seeds because the loss of a furry pelt made that holding more difficult. They would have initiated a new system of communication based on gestures, touch and sounds to prevent the babies from feeling abandoned, and the babies began to make increasingly elaborate sounds to communicate with their mothers. Perhaps, among all the myriad theories, this is the one that best allows us to understand, if not the origin of syntax, then at least the inception of the mother-ese we know today: "goo-goo ga-ga".

According to Herodotus, the father of historians as well as of anthropologists and journalists, the Egyptians considered themselves the most ancient people in the world. That was until Psamtik I seized the throne in the 6th century BC and wanted to find out who the

Egyptians really were. The answer lay, he decided, in the first word that would be spoken by children untouched by any adult influence. So the account goes in *Histories*, although Herodotus himself did not seem to give it much credence:

> ... he took two newborn children of the common people and gave them to a shepherd to bring up among his flocks. He gave instructions that no one was to speak a word in their hearing; they were to stay by themselves in a lonely hut, and in due time the shepherd was to bring goats and give the children their milk and do everything else necessary. Psammetichus did this, and gave these instructions, because he wanted to hear what speech would first come from the children, when they were past the age of indistinct babbling.

After two years, the experiment yielded results, with the children finally vocalizing something, which they went on repeating for several days. When the shepherd noticed, he brought the children before the king, who heard them say "bekos". After some enquiry, Psamtik discovered that "bekos" was the Phrygian word for "bread", and his subjects had no choice but to accept something that was not true: "Reasoning from this, the Egyptians acknowledged that the Phrygians were older than they."

4

Waiting for rain

The water in your body once flowed down the Nile, fell as monsoon rain onto India, and swirled around the Pacific.

LEWIS DARTNELL,
ORIGINS

The majority of all hunters, fishermen, and rainfall farmers who preserved the traditional way of life were reduced to insignificance, if they were not completely annihilated.

KARL WITTFOGEL,
ORIENTAL DESPOTISMS

Where crops are gods, tillage is worship.

FELIPE FERNÁNDEZ-ARMESTO,
CIVILIZATIONS

In early summer La Mancha turns yellow, because bread there is sacred. Until just a few years ago I was the only person in my village who couldn't eat bread, even if I would still kiss a loaf like everyone else did if I ever fell over. Though coeliac disease has been recognized for some 2,500 years, until very recently there has been general ignorance around it, even in hospital settings. This was why people thought I would never make it to my first birthday, and probably thought their suspicion confirmed when they were told that I would never be able to eat (their) bread. Gluten is nothing more than a protein found in wheat, barley and rye, and occasionally in oats, that damages the digestive system of coeliacs and gluten-sensitive and gluten-intolerant people, but it is present in foods and beverages such as the bread and beer that have traditionally played a central role in food and leisure throughout the Middle East and Europe, and spread later on to the United States and Australia. My diagnosis reveals that even if wheat allowed my ancestors to stay in the place they wished to be, for me personally Asia or Latin America—where rice and maize predominate—would have been better birthplaces. Although only one per cent of the world's population is diagnosed with coeliac disease, this is the tip of a rapidly growing iceberg, especially in countries that have traditionally placed gluten at the centre of their diets. Denial of gluten intolerance is on the rise, and I cannot pass up the opportunity to denounce it here.

Every time this protein is claimed not to exist, or to be a biological weapon of social control, or part of a fad, or is trivialized with unscientific claims about gluten-free diets being about weight loss, or is mistaken for lactose or sugar, the health of millions of people is at stake. Once again, we have to go a long way back in history, to the point where we left off in the last chapter, to understand why we coeliacs are still such an oddity in certain places just because our bodies reject the protein of the very cereals without which our societies would not know how to survive. Here thirst can also be found lurking.

Mesopotamia and ancient Egypt must have had their fair share of coeliacs—the individuals in question might simply never have known about the condition, and might even have died of malnourishment, anaemia or depression without ever realizing that what seemed to be keeping them alive was in fact slowly killing them. When we think of Egypt we picture pyramids and people in wigs and kohl, and not so much what that place truly was: a bread and beer factory situated between desert and river, where livestock also played a historically undervalued role. Herodotus's story about the first word may be apocryphal, but it reveals bread to be important enough that someone could think it responsible for engendering human speech.

The Ice Age came to an end approximately 20,000 years ago. In line with rising temperatures and humidity levels, the meltwater from the shrinking icecaps formed mountains and valleys and poured down into previously dried-out lakes and rivers. Deserts diminished, forests started to creep outwards. Our ancestors, who had been through three significant glaciations since the departure from Africa, went on hunting in the same way they had for more than two million years. The retreating ice gave settlers in Eurasia access to larger and larger animals, such as mammoths, bison and wild horses. They went on moving both north and south, following the migratory routes of their quarries. But the human population continued to increase, while the megafauna on which they fed came close to extinction. Something had to give.

When the waters rose again, certain places, such as the Americas, were cut off from the rest of the world for millennia. In the north of the continent not discovered by Christopher Columbus, in Agassiz, in modern-day British Columbia, the meltwater gave rise to an immense lake the size of Spain. The lake periodically gained and lost water, which, given its size, could alter the climates of far-off places despite the isolation of the continent to which it belonged.

After the ice retreated, the warming trend continued. Far away, people built small settlements in the Fertile Crescent, where Israel, Jordan, Lebanon, Palestine, Syria, Iraq, Iran, Kuwait, south-eastern Turkey and Egypt are today. It was a good land, the climate there favouring the growth of grasses, forebears of the cereals we know today, which adapted to the scarcity and unpredictability of rainfall.

The people there lived off the wheat and barley that grew wild, as well as the gazelles they hunted and the fruit and nuts they picked when the need arose. These grains turned out to be easy to store and long-lasting. By stockpiling them, they could guarantee their subsistence in times when game was hard to come by. At some point, those living in the Fertile Crescent learned that by grinding said grains, bread could also be made. They fashioned tools—such as flint-headed sickles and stone pestles and mortars—for harvesting, grinding and working with this particular foodstuff. Though still hunter-gatherers, they established permanent dwellings, and soon founded a settlement that grew into a city, Jericho, which still exists today.

All indications suggest that a day came when it rained on one of the vessels in which barley was being kept. Perhaps some curious individual decided to take a sip of the resultant concoction, and thereby experienced pre-history's second great Eureka moment. It may be that, thanks to the rain, *Homo sapiens* discovered beer by accident, just as their ancestor *Homo erectus* is thought to have discovered fire. They also still found time to make erotic figurines and befriend grey wolves, which evolved to become dogs, initially helping on hunts and, later, guarding food stores from other animals, and attaining such importance that they were sometimes buried with their owners. Dorothy Garrod, the English archaeologist who coined the name for this culture following excavations she led at Es-Skhul and El-Wad in Palestine, called them Natufians.

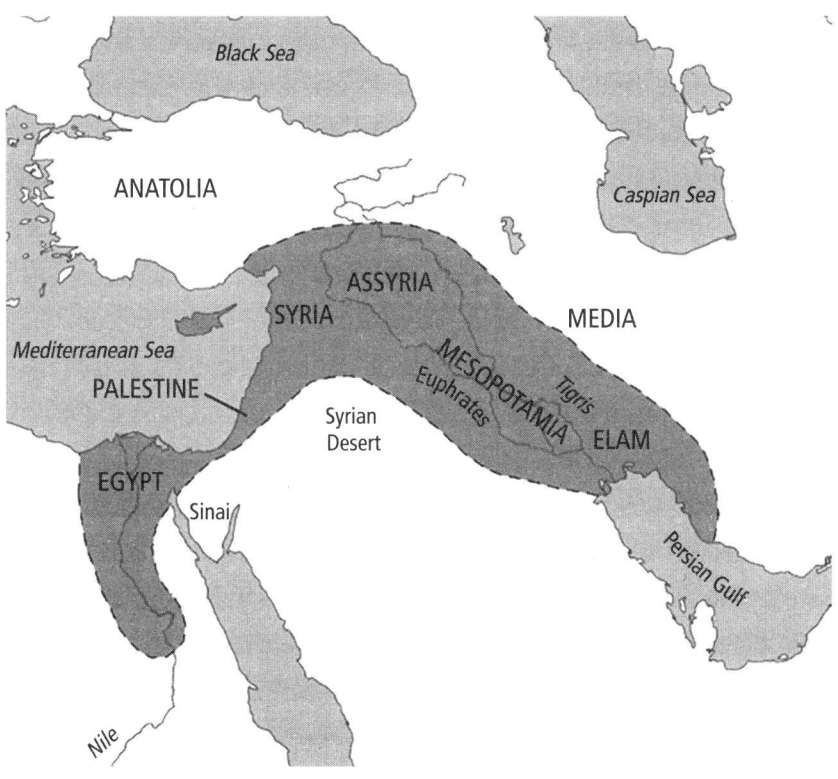

Map of Fertile Crescent

In spite of their many innovations, they would never have known anything about the North American lake that was about to change their lives. The Edenic bonanza was blown away when white flowers arrived, covering much of the world like a blanket of snow. This was the *Dryas octopetala*, the eight-petalled Mountain Avens, or Pyrenean tea. It grows in cold, dry places where snow melts quickly and gave its name to the Younger Dryas era, which came in a little under 13,000 years ago.

Between the Pleistocene (with all its drastic changes) and the Holocene (more stable in its early phase), changes in climate occurred and, although the latter is a warming phase and in turn part of a more extended cooling, there were also hiatuses, such as the Younger Dryas. The transition from one geological stage to the next was neither sudden nor gradual, but the interlude between them does seem to have happened abruptly. *Dryas octopetala* replaced virtually all vegetation in Europe for a time, leading palynologists, who study pollen and spores, to assume that the return of cold and drought was swift.

The causes of the Younger Dryas remain unclear but if we go with the most widely accepted hypothesis and rule out the possibility of a comet impact (an idea that is regaining traction) or a volcanic eruption, we are led back to Lake Agassiz. Although it emptied and refilled periodically, a crucial outflow occurred in a moment when the lake's natural dam broke. Vast quantities of fresh, cold water flowed into the Atlantic. The collision of fresh with salt water, of cold with warm, altered the thermohaline circulation, which depends on the balance of temperature and salinity and can produce changes in the climates of far-off places; in this case, as far-off as the Mediterranean Levant, where Dorothy Garrod's Natufians lived. While in Norway the cold was laying waste to forests, in the Middle East the real agent of change was aridity. Even so, it remained one of the few habitable places on Earth. So the climate, while cooling and drying much of the planet, pushed people from north to south, where at least wild seeds and grains were still plentiful. But this was soon to change as well.

The shift in climate went on for almost a millennium and meant that at some point cereals became scarce. The Natufians, it appears, were forced back to a nomadic lifestyle. But even then they did not give up their villages altogether. Although having abandoned sedentism, they would still return to their former dwelling places to bury their dead. Was the crisis of cold and thirst responsible for their discovering a sense of rootedness for the first time? What does seem clear is that they went back to sedentism once the cold and drought subsided, ushering in the first properly agricultural societies. But some researchers believe that these extremes did not push all Natufians back to nomadism. It may be that, while some abandoned their villages, others tried to hold on. Perhaps a few seeds were dropped or discarded as they were being carried to their huts, and where they landed wheat and barley sprung up. Discovering this cause-effect relationship must have been quite an event, because if they could grow wheat and barley themselves, they would no longer have to depend on changes in climate for bread and beer. Or so it seemed. On top of that, this bounty began to attract wild goats and mouflon, and game also flocked nearer to the homestead, with the result that people needed to go out on hunting expeditions only in times of scarcity.

The combination seemed ideal, but the work would be Herculean if they agreed to the pact offered by the plants and animals. The Natufians were not exactly farmers, but they may have invented agriculture, either by accident or in response to a crisis of climate, food and population, thousands of years before the Neolithic revolution. They left the land ready for their descendants. It was a transitional society, marking the change from a former way of life. But then their orchard became a steppe and their attempts to cultivate the land did not endure. They, not mitochondrial Eve or chromosomal Adam, lived in the Garden of Eden until aridity returned. Right there, in a place that is at the time of writing again suffering the effects of severe droughts, the Garden of Eden myth arose several thousand years later. What if that garden did exist and the story concerned the Natufians?

The white flower began to disappear, ending a glacial and with it a period of drought. The climate became tolerable again, marking the beginning of the Holocene, which means "entirely new time". We have been in this time now—no longer quite so new—for some 11,700 years. Although at first this warming phase promised climatic stability compared to the harshness of the Pleistocene, it has included several sub-phases in which the climate has varied, especially during the last 5,000 years when the human footprint has been crucial. Later we will see how these changes in climate have coincided with social changes. At the outset of the Holocene, with the climate more temperate but also markedly drier, someone took up an old idea and initiated a new era for humankind.

The variability of the climate during the last glaciation showed us that food security was far from a given. Added to which, humans continued to proliferate and animals to dwindle. Food scarcity and demographic pressure called for a solution that would enable the survival of all peoples, by now being *Homo sapiens* alone: with their ongoing dispersal across the globe, all other human species had been gradually disappearing, just like the megafauna. The nomadic way of life, with its system of hunter-gathering (including fishing and shellfish gathering), became no longer viable, demanding as it did that human groups limit themselves to the amount of meat available in their respective areas.

It was a vicious cycle to which humans appeared condemned; with enough food, the population grew, but before long there would be too many people to feed. Back to square one. We have already seen that primates initially fed on insects, but when fruits began to appear, they began eating those. It fell to *Australopithecus* to do something about it when succulent fruits became scarce. A change of diet was their only option, and fortunately evolution had endowed them with the necessary anatomical and morphological adaptations to eat what was at hand. So they now added insects to their diet, the capacity to digest which they had inherited. Later on, *Homo habilis* subsisted on seeds, fruits and roots, but they would also hunt the occasional small animals. *Homo erectus*, on the other hand, had to go all in for meat, doubtless carrion to begin with and only later, larger hunted animals. All the signs suggest roasted meat to have been their preference, and after expanding throughout Eurasia they passed down such lifestyle choices and diet to those who came afterwards.

But when such scarcity and demographic pressure had previously coincided, *Homo sapiens* left Africa and spent almost the entire ensuing Ice Age on the move. The mass migration route was exhausted because during the Younger Dryas the migrants had already spread and multiplied on every single continent. Now meat was in short supply. A change in diet did not appear the best option. But neither was it the worst, so they increased their consumption of wild cereals, which, having adapted to drought conditions in the Fertile Crescent, still grew readily. Although the increase in cereals over meat resulted in a lower nutrient and calorie intake, the deficit could be reversed if they took charge of growing the cereals themselves; the amount of food eaten could be increased to compensate.

This idea of cultivating the soil, which was not entirely new, must have created some serious dilemmas. It was far from a cure-all: it tied people down in a certain place, imposing tasks that only multiplied in time, and made them entirely dependent on the rain, given that irrigation techniques had not yet been developed. The conditions were punishing, but the general view is that they accepted the contract as something that would enable the maintenance of a large population, which in the new circumstances would grow even more. Cereal-based, non-irrigated, subsistence agriculture emerged as the solution to all their woes, and was taken back to the Fertile Crescent, where it remained for good. It promised food for all, but also the capacity to generate surpluses that were later key in the formation of cities and states.

The global event termed the "Neolithic revolution" by archaeologist Gordon Childe in fact took place independently in geographically very distinct places only a few millennia apart, and in some of them virtually simultaneously. It soon entailed a new way of life for the peoples of Eurasia, most of whom left the hunter-gatherer way of life behind definitively. Save for a few exceptions (still extant in certain places), *Homo sapiens* ceased their wanderings and turned farmers.

People have at times been reluctant to believe that hunter-gatherers who had lived long distances away from them could possibly have come to the same conclusions as their own, geographically proximate ancestors—in their eyes, the only truly civilized ones. What's more, such an idea diverged from the notion that human beings had developed in intelligence, knowledge and overall sophistication precisely because of agriculture and all its ripple effects, which had taken them away from the wildness of nomadic peoples. The idea also seemed to pass over the importance of livestock farming and fishing. But we now know that geographically isolated peoples, separated by tens of thousands of kilometres, began seeing the same possibilities in a very short space of time without any enlightened people coming to instruct them, and they put them into practice with whatever crops grew best in their respective lands. In China, for example, a little over 1,000 years later, rice and millet began to be cultivated in response to arid conditions. Pigs were also domesticated. Almost at the same time, bananas and cane sugar were first cultivated in New Guinea. In Mesoamerica someone gave corn- and bean-growing a go, while in South America llamas and potatoes led the way. West Africa, which had its own varieties of rice, wheat and millet as well as sorghum, focused on these crops just over 5,000 years ago. At that point, people in the Fertile Crescent went on working with what was to hand, and so olive trees and grape vines, and cattle and pigs, came to the fore.

It matters little whether we are omnivores, ovo-lacto-vegetarians, vegans or "realfooders", if we eat ultra-processed foods or if our bodies have imposed a gluten- or lactose-free diet on us. What we are (primates) and almost everything that provides us with what we eat (artiodactyls such as pigs, cows, sheep and goats, and angiosperms, which give us cereals, legumes, fruit and vegetables) first appeared in the world around the same time, when the dinosaurs had given way to primates and the Earth was cooling and drying. We have the domestication efforts of our ancestors during a window of a few millennia to thank for most of the plants and animals on which we still feed ourselves. And all these foods were place-specific, in the Fertile Crescent, China, Mesoamerica, the Andes and New Guinea, until they were transported elsewhere.

But why was the Fertile Crescent the first place where this happened, albeit only slightly earlier, and why did the practices spread there more quickly? For one, it was especially rich in plants as well as domesticable animals, even if gazelle numbers were on the wane, making hunting less viable. At the same time, certain indigenous cereals went through a moment of expansion alongside human population growth. Tools and implements had also been invented in just this area that favoured the production of fast-growing foodstuffs requiring relatively little tending. Jared Diamond looked at the map and found another possible explanation that linked the layout of the continents to the rate at which agriculture took off. He thought

Eurasia's horizontal spread and lack of natural barriers, as well as climatic and ecological similarities, favoured the east-west dispersal of agricultural practices, while in the Americas and Africa there were more obstacles along the north-south axis.

Now that we know about the various attempts before agriculture's starting pistol was fired, it seems impossible not to look at another map, that of *Homo sapiens*'s global dispersal, and think that something else could have happened: what if they had inherited this knowledge or the spark of it from their ancestors? The emergence of agriculture in the Fertile Crescent does not seem to be a coincidence, especially in view of what happened in 1989. The drought with which this book began was also wracking other places in the world, such as Israel, and when the waters of the Sea of Galilee receded, they revealed things that had been hidden for thousands of years. The remains of a settlement could now be seen, with its huts in which grain had been ground, and where sickles and mortars were used. There were also early signs of cereal cultivation from 23,000 years ago, at the time when the Ice Age was in its death throes. All of which was long before the date considered to mark the onset of agriculture and even before the Natufians brought bread and beer to the same place.

Furthermore, before all of this, some 70,000 years ago and in that very place, Neanderthals had clearly known how to make a kind of unleavened bread. Various Sapiens from Africa made a long stopover there, some of them staying on, while their descendants had made several attempts to cultivate cereals, until finally succeeding tens of thousands of years later. What if it was Neanderthals or one of our African grandmothers who gave us the recipe for bread and beer, or even the keys to the cultivation of cereals? The descendants of those who had dispersed around the world came to the same conclusions at almost exactly the same time. Was there a shared innate ability that led them to these conclusions, or had they instead learned something before they all set off on their global peregrinations?

The question of whether agriculture or sedentism came first is almost as old as that of the chicken and egg, but a long-standing consensus on the order paved the way for a narrative in support of state policies that favour and prop up agriculture. The idea, broadly, was that agriculture was invented almost 12,000 years ago in the Middle East, and that this gave rise to sedentary ways of life, and this in turn to cities, with their monumental buildings and fortified walls, and the state, and waterworks, and the written word. Civilization had begun. When the origin of civilization is explained, it is usually with recourse to ideas about progress, with the first agrarian states and thus agriculture figuring as both starting point and exemplar.

Within this narrative, those poor ragged *Homo sapiens* were exhausted after stumbling about in search of food and shelter, until they finally invented agriculture and, like the heroes of their first myths, both gained access to the grain of the gods and saved mankind. They were the chosen few. They were now tied to a place, obliged to work from sun up to sundown, but this at least meant they had a home in which to lay their heads, and the attendant peace and stability made for a major improvement on their meagre existences to date, in turn enabling them to develop their minds and the arts. The state then came into being, and people could finally be counted as civilized (in contrast to the hunter-gatherer nomads of before). Although there was now no need for them to remain on the move, they set off to the four corners of the Earth to explain to everyone else how life had surely been until then. The natives were wowed, and fell in behind without a second thought. They shut themselves up first inside their homes and then in their cities with no hint of homesickness whatsoever, because that was the way to progress in life. They asked the plants and animals to please give them more work, and the enlightened ones to instruct them in the best way to break their backs building irrigation channels and ditches. Some sense of irony is all it takes to realize that this was hardly the case. But, although exaggerated here, this narrative has none the less been dominant and in a way still is, since it also served the later dispensation so well, even though it subsumed and marginalized rural folk.

Life could not have been easy before people turned to farming, but equally it is a stretch to believe that they would have chosen to exchange a few hours a week hunting and gathering for 24/7, land-based toil. Although agriculture came in gradually, and seems at moments to have fallen away, it was likely an ordeal, if not for the very first adopters, then certainly for the next wave; the earth and the crops becoming accustomed to human intervention actually increased demands. Tilling, fertilizing, sowing, waiting for rain, weeding, harvesting. And then start all over again. We have no idea what their conception of leisure was, but it seems no overstatement to suggest that previously, with just three weeks gathering wild wheat, they would have had enough to feed their family for the entire year. That made for forty-nine weeks of holiday in a year, interrupted only by the moments when they needed to go out and hunt. Had they really developed Stockholm Syndrome before being locked away? Would they really have jumped at the prospect of hard, all-day physical labour? This is the same narrative that today encourages people to take cold showers to endure eighteen-hour workdays. Promoting hyper-productive strategies to support subhuman working conditions while advocating a return to a paleo diet—this perhaps is the final great bundle of contradictions created by a discourse that has been around for thousands of years. But if society did not tell itself this story, it could no longer tolerate the idea of living to "earn a crust", and doing so in a state of constant fatigue.

Although there is no way of knowing what these people thought or felt, the archaeological, genetic and palaeoclimatological evidence suggests at least that the order of things was not exactly as it has been portrayed. Dorothy Garrod's discovery, the subsequent findings about the Natufians, the drought that decades later revealed the secrets at the bottom of the Sea of Galilee, and even the remains of the pancake eaten by a Neanderthal combine to reverse the accepted order and invite us to rethink whether Natufian culture was really just a failed experiment. What if, rather than building houses in which to store grain, they discovered that such things already existed? We also cannot know when their first settlements came about, because the materials they used may have vanished altogether, even if some evidence of a wooden construction from 500,000 years ago in Gambia has lasted. This is why it is difficult to know for certain if we are nomads by nature and were forced into sedentism or vice versa; whether bread is saviour or tyrant. Did agriculture take us hostage, or provide us with the peace and quiet necessary to innovate and create? Was it all the result of hunger and thirst, or of plenty?

Before the advent of agriculture, there were not only settlements, art, bread and beer; the first recorded temple had also already been erected. It is not clear who built Göbekli Tepe ("pot-bellied hill" in Turkish) or for what purpose, but it seems to have been the result of the collective effort of thousands of hunter-gatherers with shared beliefs, because there is no evidence either that it was attached to any kind of settlement. Nor is it known why this monumental construction was built underground. Göbekli Tepe, which predates Stonehenge by thousands of years, may well have been the place where the notion of the sacred first emerged, and the possibility is now being raised, not without its opponents, that rather than the agrarian state *this* was civilization's true trigger. The principal figures on this temple's carven pillars are animals, bovids especially. Something very similar must have happened nearby at Çatal Höyük, one of many places apparently abandoned in the middle of a drought and that also has depictions of bulls' heads and horns on its walls.

The pro-agrarian state narrative neglects certain significant facts, such as that agriculture arose elsewhere independently and, as anthropologist James Scott reminds us, thousands of years had to pass between the emergence of agriculture and sedentism (in the Americas) and thousands more between sedentism and the state (in the Middle East). It also did not always happen in that order, and there are nomadic groups still in existence today that are fighting against the imposition of a sedentary lifestyle, and indeed the stigmatization of their traditional way of life. These omissions seem far from innocent. The same goes for a conversation that has been circulating on the internet for some years. According to this story, anthropologist Margaret Mead, when asked for her thoughts on the origin of civilization, replied that a broken femur found at an archaeological dig, which showed signs of having been tended and healed some 10,000 years ago, marked the origin of civilization. The account is too brief to be reliable, the date too close to our present moment not to be doubted, and I have been unable to corroborate the story or know to which site she was referring. I did find an interview in which Mead answered the same question, but in a completely different way: she spoke of city-building and a new regard for posterity, signalled by the advent of the written word.

Mead wasn't alive when archaeologists at the Atapuerca dig in Spain came to one of the great conclusions about their findings at Sima de los Huesos: if at least one old man, one sick man and one dependent girl had lived together in that place, there could only be one reason why they lived as long as they did. They would all have died far sooner without the care of the group. Four hundred and thirty thousand years ago, *Homo heidelbergensis*, descendants of *Homo antecessor* and ancestors of *Homo neanderthalensis* who lived at Atapuerca—long, long before Mead's hypothetical femur—already knew what it was to heal and care for others. And humans may have always known this, because even chimpanzees make natural poultices with insects, and we know of mammals that have adopted abandoned or orphaned offspring belonging to other species. The viral story, in which Mead likely did not feature, when coupled with reports of supposedly violent, cannibalistic and isolated tribes, who not coincidentally maintain a life based on hunting and gathering, helps to reinforce the already widespread narrative that civilization was sparked by the domestication of animals and plants.

While some believe that the decision to cultivate cereals was the first link in a chain of progress, others think that this was a case of humans expelling themselves from the Garden of Eden with their greatest ever misstep. I am not advocating a return to Palaeolithic life, for violence and inequality were surely already a reality then, nor do I idealize the life of the Flintstones, but I do think it is important to stress that the hunter-gatherers who still share the world with us were not "left behind", because, among other things, they actively chose to maintain that way of life and because their needs have long since ceased to be our needs. If some of them are still isolated, it is because proximity with us is deadly to them. We, on the other hand, got ourselves into a bottleneck that simultaneously created more and more needs while making us increasingly vulnerable to nature as well as increasingly water-dependent. If we make the effort to put ourselves in the shoes of a hunter-gatherer beset by hunger, thirst and cold, who has to change her way of life and start cultivating the land, it isn't actually a dream solution that appears before her, at least in terms of nutrition, health, work and leisure.

If we are going to dwell now, above all, on the Middle East, it is because it was there that wheat and barley were first cultivated, and because from there both were taken to the place where this book began—by Anatolian farmers and steppe-dwelling shepherds who had a significant impact on La Mancha.

The bones of a whale that had died there tens of thousands of years before were found in the world's largest desert. When the Tethys Ocean dried out and contracted some seven million years ago, forests, savannahs, lakes and areas of swampland eventually flourished. Some of the lakes were larger than certain present-day seas and the water filled great basins that are nowadays covered in sand. Mammals that succeeded in becoming water creatures again, as well as the largest animals the world has ever known, originated in the Sahara—although no one would have called it that at the time, given that *sahra* means "desert" in Arabic. It is an area that has periodically dried out and become green again ever since.

Rain and drought were seasonal, meaning that the Sahara and the Nile alternated in attracting and repelling the peoples who had lived there for generations. Their ancestors sought to be near the river for access to game; never too close, given how high it rose when it flooded, but never too far away either, since in the other direction all was desert. A reduction in Saharan dust altered the rainfall pattern. Since the rains turned torrential and the desert turned green again, some people who had until then lived halfway between the Nile and the Sahara made their homes around natural oases. Only during the dry season did they have to move elsewhere, always returning home with the rains. That trend was reversed when the Holocene Climate Optimum ended around 5,500 years ago. Both the levels of dust and sunshine increased again, the rain stopped falling, and humans had to abandon those oases, now swallowed up by desert. In the eyes of some researchers, this was an abrupt process, over in less than 200 years. But others think the transition was actually rather gradual, with the Sahara drying out and becoming a dustbowl over a stretch of 3–4,000 years. Although it has always been attributed to periodic changes in the Earth's rotation, debate is ongoing over the extent to which human influence has combined with astronomy to alter the climate and reduce the rainfall that once brought monsoon winds to North Africa.

What had for two millennia been pleasant climes were suddenly overtaken by a mega-drought, with knock-on effects elsewhere in the world. Desperately searching for water, people relocated in droves to springs and rivers. Apart from the Nile, only the Tigris, Euphrates, Yangtze, Yellow and Indus rivers were still flowing at a good rate. Those climate refugees settled on their banks and began clustering together—in small groups at first, though these eventually grew and gave rise to the first ever civilizations.

The peoples living relatively near the Nile Delta had occasionally to return to a river that provided them with food, but was terrifying when it swelled. It drew shepherds from as far away as Libya and Numidia, and later on Hamitic tribes (from what is now Ethiopia). Thirsty, spat out by the desert, many found a habitable region midway between the water and the sand. They began assembling in "nomes", small plots that, if irrigated, could be cultivated, and around which their first villages emerged. They learned to live with, indeed organized their new culture around the cyclical flooding of the river. Such was their mastery of water that 4,800 years ago they had already built the world's first dam.

The villages grew and inevitably began joining forces with neighbouring villages, and evolved into cities, which in 3200 BC saw the kingdoms of the north and south come into being. Ancient Egypt was now a reality. They called their country Kemet, which means "black land" or "fertile Nile mud that prolongs life", its name inspired by the colour of the silt that covered everything during the valley's annual flood. But Kemet also distinguished it from *Deshret*, which meant "red earth"—sand, that is—and was used to designate both "desert" and "foreign land". Hence perhaps the death-symbolism of the colour red for them, and the association of black with regeneration. In their cosmovision, a mythological heron named Bennu was the harbinger of the rising waters—which always coincided with the return of the herons.

It was not long before desert and river had their own gods. And not only in Egypt. When human beings began to work the soil, they ceased depending on indigenous seeds and grains, ushering in a new relationship of dependence on the heavens. They were compelled to come up with rain-providing gods, to explain both where it came from and why it sometimes didn't come at all, and to know whom to ask for it. And where else were such beings going to live?

In ancient times, various cultures believed that the world was a kind of disc topped by a blue dome in which the gods resided. Hence the idea that some of the water was suspended in the sky, and that it sometimes fell to earth. In fact, the first Egyptians thought that rain fell from a celestial Nile, and in Mesopotamia it was believed that Tiamat's tears had produced the rivers. The stars were conceived of as divine beings who in certain cases travelled by boat and could send rain and storms, or choose to withhold them. This perhaps explains the belief among the Sumerians, Egyptians and Mixtec that, respectively, Enki, Hapi and Dzahui were in the habit of emptying jugs of water from on high.

In Mesopotamia, Egypt and China, the story was soon fleshed out more fantastically still. While in Egypt Seth and Osiris competed to bring death or abundance, China had its equivalent in the myth of the god of war Chiyou and the Emperor Huang Di, who used his daughter Nüba, goddess of drought, as a weapon to overthrow the former. Other versions and myths explain the origin of Chinese civilization as stemming from a major drought. The first tells how Tang, a conqueror king, went to the mulberry forest to make a sacred request: he offered himself as a sacrifice to end the drought that was afflicting his people. Another version, however, tells of a woman dressed in green—symbolizing regeneration—climbing a hill and lying down to let herself dry out in the sun.

But to return to Egypt—where the city of Oxyrhynchus gets its name from the Hellenized version of other monikers honouring the medjed, a species of elephantfish long venerated on the banks of the Nile. Though the medjed's merits might appear questionable, it did perhaps save Egypt from going thirsty—in the myths, at least. Varying versions posited Seth and Osiris as brothers, the former inheriting the desert, the latter the fertile lands. Curiously, and through no etymological relationship, a being called Seth became the god of both drought and the desert. Seth had the body of a man, the head of a dog and was able to turn into a snake, although in some depictions he looks more akin to a hippo. He was often compared to a pig, which was a way of discrediting an animal that people had by now begun to view with suspicion, and he was associated with the death-symbolizing colour red. But Seth was not always feared and hated, and Apopi I made him the sole god of the temple of Avaris. Unsurprisingly, a civilization founded by water-seeking nomads created a red-hued, chaos-bringing god of thirst embodying all of life's evils; correspondingly, his nemesis was a god of fertile land associated with the colour black. With the story of Seth and Osiris, the Egyptians created their own Cain and Abel long before the Bible.

In some myths Seth drowns Osiris, shuts him up in a tailor-made trunk and chops his body into pieces. Others seek to explain why an enraged Seth killed and then chopped his brother up. But whether or not Osiris slept with his sister-in-law, as some versions have it, he was always envious of his brother, he of the fertile lands, he who nourished the clouds, the saviour who introduced Egyptians to the wonders of agriculture. Seth scattered the parts of Osiris's body far and wide, leaving Isis, his widow and sister, with the task of putting him back together again. According to one version, her dead husband's phallus was finally returned to her by a fish that had swallowed it whole—a medjed. Isis was then involved in one of the great acts of divine necrophilia when she managed to conceive by Osiris, later giving birth to Horus; though in another myth, Osiris briefly comes back to life, joins with Isis to engender Horus and then departs to rule the underworld, and it is then that the heron Bennu becomes Osiris's "ba", the personality component of the soul.

All of these versions in fact speak to the Nile's periodic rise and fall, and the antagonism of thirst and abundance. The inundations gave origin and structure to the Egyptian calendar. Owing its existence to the constant cycle as that civilization did, it divided its year into three seasons: firstly, the flood; secondly, sowing and germination; thirdly, the harvest. It also came up with the nilometer to measure the river's water level, as the wheat and barley crops depended on this. But, although they believed that Hapi warned the priests that he was going to empty water from a jug up in the heavens, it was only with rainfall in the mountains of Ethiopia and the river's annual thaw that everything would flood, bringing new life to Egypt. But then came a time when it didn't rain in northern Africa, and the person charged with marking the water levels using the Palermo Stone had to indicate that the river had barely risen.

Just as thirsty nomads were coming together on the banks of the Nile, small groups of people, possibly no more than ten in each to begin with, were creating settlements between the Tigris and the Euphrates. Nobody knows for certain where they came from or why, although displacement from both deserts and mountains by the drought at the end of the Holocene Climate Optimum is feasible. They founded small agricultural hamlets, in the environs of natural springs initially. Then they began drawing nearer to the rivers in northern Mesopotamia, which means "land between rivers", and spread in the intermediate zones and finally southwards. It was in the south that Sumer arose, which means "the land of the lord of the reed beds". Considered the first civilization in history, it developed in what is modern-day Iraq, although the most likely thing is that both the Sumerians and the first Egyptians settled by their respective rivers more or less simultaneously.

The need to manage the water led these tiny hamlets to develop into large cities, although this left the population increasingly vulnerable. It was no small feat to set up in places constantly subjected either to flooding or prolonged droughts. They had to deal with the increasing water salinity, deforestation—their own doing—and demographic pressure. Those small groups mushroomed to tens of thousands, and the cities, although small by today's standards, were enormous at the time, and not always viable. Clashes over management of both water and the fertile areas of land grew in intensity, until finally the first war on record broke out.

First the Akkadians, then the Assyrians and the Babylonians followed the Sumerians in settling Mesopotamia. Their contributions to our world were considerable: they laid the foundations for irrigated agriculture as well as for the written word and astronomy, and the plough and the wheel were also their inventions. The first known attempts at subterranean hydraulic engineering were undertaken in Babylon, in the form of the *qanats* that may have been taken elsewhere by Arab peoples. King Ur-Namu was credited with having "built" irrigation channels through the middle of the desert, connecting the city of Ur with the Euphrates, although he, in an act of false modesty, said it was all down to the gods. But Mesopotamian kings were precisely a kind of incarnation and messenger of the gods; attributed with the power of invoking the rain, they also took responsibility if their petitions fell on deaf celestial ears.

The settlements in this time were mainly concentrated in the semicircle of the Tigris, Euphrates and Nile. The Fertile Crescent was one of the few habitable places in the world, but not the only one. When there is talk of the first civilizations, one can have the impression that nothing was going on beyond the borders of cities. Aside from all the nomadic peoples who remained on the other side of their walls, the Caral culture arose in what is now Peru, contemporary with Sumer and ancient Egypt. There, although nearly three millennia later, the Moches built an extensive, complex system of channels with which they succeeded in irrigating their desert lands with water from distant Andean rivers. On the Iberian peninsula, not long after Sumer and Egypt were founded, El Argar also arose, which was home to the first urban society on the Peninsula and perhaps in Europe, in present-day Murcia, Almería, Granada and Málaga. El Argar went out of existence in the middle of a drought, leaving questions about the relationship between its collapse and the climate—there are signs of both deforestation and constant fires—and also leaving the oldest example of the object perhaps most useful in and typical of dry Iberia: the water jug. It is believed to be a local invention, the fruit of our ancestors' ingenuity, though an older version was already in use in Mesopotamia.

El Argar was also coeval with the "motillas" settlements of La Mancha; one of history's great unknowns, it has been suggested that these settlements formed part of Europe's first ever hydraulic society; we will return to them in the next chapter. Very soon after the founding of Sumer, or possibly at the same time, the building blocks of Chinese civilization were set down on the banks of the Yangtze (where the Liangzhu culture arose) and the Yellow River (the Xia and Shang dynasties). In the Indus Valley, in what is now Pakistan, the Harappa civilization came into being. Although Mesopotamia or Egypt appears to be the origin of it all, it matters little which was the very first: the point is that virtually simultaneously, in various places in the world, different societies came about with water as their organizing principle. In all of them, cities with some kind of water-management system were built, and proto-writing also developed. It has been suggested that this was when humankind first began flirting with social inequality. All established in the vicinity of rivers or on floodplains, these cities attained greater levels of complexity when they sought to divert and harness the water and, above all, when they began to succeed in doing so to the best of their abilities using irrigation channels, dykes and canals. Once they had created stable sustenance for themselves, some of them were able to develop the proto-scientific and cultural frameworks for both the written word and technologies that have gone on to be decisive over the millennia.

Karl Wittfogel, a writer now little discussed, called them "hydraulic cultures" for the fact they had river water and large-scale channelling works in common, as well as that these served to reinforce the power of the ruling classes, whom he called "hydraulic despots". The biologist Lewis Dartnell has found another common component that relates to thirst and how our desperate search for water makes us especially vulnerable. As he says in *Origins*, most of these early civilizations, China and Egypt aside, and many later ones too, were built along the edges of tectonic plates unbeknownst to them—their peoples had no idea that setting up on fault lines dotted with so many springs also exposed them to earthquakes, volcanic eruptions and tsunamis. But it seems far from coincidental, given the access to water and fertile land these places provide.

Just as hydraulic civilizations were coming into being, drought returned. In 3800 BC a shift in the Indian Ocean's monsoon corridor meant that the rains arrived later and were less heavy in Mesopotamia. In 2450 BC, the oscillation of the Earth once more modified the amount of sunlight it received. Changes in solar activity altered the rain patterns, and several arid centuries ensued. Agricultural expansion had led to deforestation, and the ravages of civilization were beginning to be felt. Volcanic activity may have further altered the climate.

At this time, different peoples came up with different answers in response to thirst. While civilizations like the Harappa were founded in the Indus Valley and the Xia were taking their first steps in what is today China, descendants of herders who had long before begun drifting away from the Pontic-Caspian steppe were spreading across much of Eurasia. Umma and Lagash clashed in Mesopotamia over an area of fertile land known as Guedena. This "water war", as they called it, is the first war of which we have any record. But, directly or indirectly, thirst has been the cause of many other conflicts besides.

At the beginning of the 1980s, someone took a metal detector with them on a walk in the countryside near Zaragoza. When the device beeped, the individual duly dug until a piece of bronze was revealed. In exchange for their identity not being revealed, the person handed over the find. The second Botorrita bronze, known as *tabula contrebiensis* and dating from the first century BC, had survived, and captures the peaceful resolution of a dispute over the waters of the Jalón River between the inhabitants of Salduie and Alaún (today Zaragoza and Alagón, respectively), with the intervention of six judges from a neutral neighbouring town. Salduie wanted to build an irrigation channel across what Alaún considered to be its lands. Although they had a similar problem to that of Umma and Lagash, they turned to a neutral actor in search of a non-violent settlement. "Since we have the power to judge, we rule in favour of the Salluians in the matter in dispute," wrote the magistrates of Contrebia Belaisca, Botorrita.

Because they arose in various places, the precise origin of writing systems is uncertain, although everything points to someone in Mesopotamia, around 5,000 years ago, coming up with cuneiform as a means for making something we still busily produce today: lists. In that case, clay tablets were used, but papyruses came to Egypt at much the same time. Water and its absence meant that this book would one day be printed. Although papyrus was first created on a riverside, it needed somewhere dry in order to last. Irene Vallejo explains that the only reason we know certain stories written thousands of years ago is because the sand, aided by the absence of rain in a dry climate, kept the papyruses intact; normal levels of rainfall would have destroyed them.

Long before paper existed, before papyrus too, and while clay was still being written on in Mesopotamia, in China turtle shells and animal bones were being etched with a protolanguage that had arisen along the Yellow River, a watercourse that gets its name from the colour of the clay carried on the winds from the Gobi desert every winter. There, river and desert merge, the latter remaining suspended in the water.

A man had to contract malaria in 1899 for the world to find out about the protolanguage contained on those "oracle bones". Field workers would take those bones to market and sell them to apothecaries, since it had been common practice for thousands of years in China to use their powder medicinally. "Dragon bones," the remedy was called. One day, with the help of a friend, the man in question was grinding some up as a cure for his malaria. Halfway through the process they noticed a difference in colour and, stopping to take a closer look, found characters inscribed that were similar to those still used in the country at the time. Early the next year expeditions began in Yinxu (near Anyang, in Henan province). More than 100,000 inscribed bones were found in this spot alone. All contained a question and answer, and some additional comment. Unlike other fledgling systems, this Neolithic Chinese proto-writing had nothing to do with accountancy, rather being concerned with predictions about the immediate future. And a few thousand years ago, one of the most common and pressing worries for the people on the banks of the Yellow River—who would seek an audience with the king in order for him to tell the future using a bone or shell—was the rain. This was reflected in one of the questions that has been recovered from the oldest of these oracle bones: "Is it going to rain today?"

Others were uncovered, their messages having been interpreted as invocations to the rain, which have been with us ever since inscriptions were first carved on bones and stones. The symbol for rain on an oracle bone from the Shang period was made up of three vertical lines separated by small gaps, practically the same as in an ancient Egyptian hieroglyph.

As a culture that arose alongside rivers, China is rich in water legends. According to one, Cang Jie, astrologer to Emperor Huang Di, had four eyes and the head of a dragon. He was rumoured to have been able to write from birth. He sought inspiration in nature to create the writings commissioned by the Son of Heaven, as the Chinese emperors were known at a time when a flood or a drought was enough to bring about a change of rulers. It was said of the emperor that his mother had been made pregnant by a bolt of lightning, that she had not given birth until twenty years later and that, when the child was born, he was able to speak from the moment of his birth. As the ancestor of the Han, he carried a drum made of the skin of a kui, a mythological being with the power to bring both rain and drought.

Cang observed the sky and birds' tracks after a spring storm. He also contemplated the shapes of turtle shells, feathers, mountains and rivers. Once his greatest creation was ready, millet began to rain from the sky, the spirits wept nightly and the dragons were never heard from again.

The myth about the origin of writing is strongly linked to the April and May rains, to the extent that the "gu-yu" pair of characters refers both to the seasonal showers and to the sowing season, as well as a day in the calendar on which Cang Jie is said to have invented Chinese ideograms: 谷雨 is 20 April, Chinese Language Day.

The Chinese were not the only ones to show their preoccupation with water when they first began communicating with symbols on bones and stones. The Codex Hammurabi contains the first known laws on the subject and we have already seen that the oldest known court case on the Iberian peninsula centred around ancient conflicts over its use.

But long before that trial, while the friezes of Caral, the Stele of the Vultures, the Palermo Stone and the oracle bones attested to the shared thirst of the Caraleños, Sumerians, Egyptians and Chinese, in a place in Andalusia that is now Zalamea la Real (Huelva), as the rain continued not to fall, someone looked to the sky. There was no sign of any change. This person went up to a rock and began to carve what might have been his wish: the rock was gradually transformed, with each impact, into a representation of the heavenly vault. He drew circles, struck harder, and made inscriptions that combined to turn the stone into something resembling a puddle. What does rainwater look like when it meets the still water of a puddle? Like the Aulagares petroglyph, concluded José Luis Escacena.

The same drought was wreaking havoc in what is now Germany. Someone came up with the idea of dropping coins into a well as a way of making wishes, of telling the heavens that the sound of coins hitting the water—the sound of rain—was what they longed for most. According to various interpretations in very distant places, it is possible to say that prehistoric art elsewhere features similar petitions, for example in the Ariquilda I petroglyphs (Chile) and those in Jalisco (Mexico), as well as in the bowls and dippers that are so numerous in Albacete and in Las Palmas de Gran Canaria, Spain. And, although these are only hypotheses, they are repeated often enough for us to see a thread that connects these thirsty people in a way that is both sad and beautiful: the memory of their thirst has been left on the stones. They did not know each other. Only now can we see what unites them. What indeed unites us all.

5

Under dry ground

> Desert is a loose term to indicate land that supports no man; whether the land can be bitted and broken to that purpose is not proven. Void of life it never is, however dry the air and villainous the soil.
>
> MARY AUSTIN,
THE LAND OF LITTLE RAIN

The hill looked like a bobble hat fallen from the sky. There in the foothills of the Sierra Morena, a scattering of white boulders at the top of the Bonete rise was sufficiently striking to earn it the name of *Castillejo*. Sometimes *castillejo* is used in La Mancha in a derogatory way (it would be *castilla* for an actual castle), in reference to the ruins of what may have been a castle, now long forgotten. In any case, the truly striking thing from up there was the view of the Peña del Cambrón, a peak in the facing Sierra Segura that looks almost like a made-to-measure sun altar. And that particular *peña* (crag or rocky outcrop) had all the ingredients for what, in the eyes of any animist, sun-worshipping society, would be considered a miracle: at the winter solstice, the sun would rise as if directly out of it.

It was no coincidence that the hat's bobble was situated on the final outcrop of the Meseta Sur plateau, on the border between the Guadiana and Guadalquivir river basins. Castillejo del Bonete overlooks a natural corridor connecting the plateau to Andalusia, worn by age-old footfall. From there, it was possible to see anyone approaching in the distance, from either south or east.

It appears that when the Romans arrived in Iberia to take on the Carthaginians during the Second Punic War, they were lulled by the temporary mirage of a climatic phase so favourable that it later came to be known as the Roman Climate Optimum, or Roman Warm Period. And although this period would not last forever, they decided to settle, above all because of the economic possibilities offered by the land here. Spotting the old route, they went on to widen and slightly alter its course so that it eventually crossed the Peninsula from south to north and linked up with other thoroughfares beyond the Pyrenees. Thus they linked Gades (Cádiz) with Rome, a journey that took around three and a half months on foot, and turned it into Roman Hispania's principal highway. It was named after the emperor who ordered the improvement works that now made transit so much easier: Octavian Augustus. Next to the Via Augusta, near modern-day Terrinches—which did not yet exist as such—a Roman whose name has been lost retired to the countryside and built himself a villa with baths, which came to be known as Ontavia, which roughly means "where the road is". The first ever counter-urbanizer's choice of location was confirmed in later times when his villa was chosen as the site of an archaeological dig.

The Via Augusta also came to be known as Hannibal's Road because, unlikely as it seems, the Carthaginian general Hannibal Barca is said to have travelled along it with his legions and elephants when preparing to cross the Pyrenees and the Alps on his way to invade Rome. Later, it became the Via dei Vasi di Vicarello when some silver urns engraved with the entire route appeared near

Rome. These are thought to have been votive offerings from a traveller, linked by some to a man from Cadiz who made the journey to Rome on foot just to meet the historian Titus Livius.

Apart from people and silver urns, the Via Augusta was also a conduit for cereals, oil, wine, wool and metals. But in the first century BC, a drought arrived that parched the crops and, three centuries later, at the peak of the aridity, the Romans started to leave Spain without so much as a goodbye—and without repealing the Edict of Caracalla either, which six years earlier had given all free men born in the empire full Roman citizenship. Did they leave on account of thirst? Some historians believe so.

Amidst all these changes and comings and goings, there in Oretania (present-day Ciudad Real, Albaceta and Jaén), the bobble hat remained just where it was. At the foot of that peak, recent generations of my family eked what water they could from a cave until thirst drove them away for good.

Our village was always awash with myths and legends concerning underground treasures, which spoke to an atavistic relationship with the subterranean world. A phrase often on the lips of the elderly, which they attributed to other villages, went: "If Terrinches people knew what lay beneath,/ they'd be down there day and night, digging with their teeth." People spoke of a sunken passageway running from below the Castillejo to someone's house, and about the caves closest to the castle in which people had lived ever since the days of Arab rule. At the foot of Castillejo del Bonete, my grandfather planted his last olive trees, and he often took me up to the caves, until recently still inhabited. I spent my childhood scratching around in that soil, but I never found any sign of my ancestors beside old medicine bottles. Though an encrypted message was passed down to us in the form of the place name, and though local shepherds retained the knowledge, it took us a long time to get at the treasure, and all the signs are that it was also missed by the passing Romans, and by Hannibal, and by those people taking urns to Rome. It was there all along.

Not that the bobble hat from the sky could have hidden it. Only the village shepherds, who had been herding their goats and sheep there for five generations, suspected that there might be something under the stones. Their grandparents had told them that it had once been a village overlooking the main road and that a cave was hidden beneath the ruins. Plus they always saw ash from fires there.

Midway through the century, a farmer yoked his cows and took them up there to till the soil, preparing to sow vetch (its flour is used in our Manchegan rainy-day pease pottage). The man often said he thought there was something about the place, given the stones he always turned up when tilling. But olive trees were then planted, partially covering it over, and the shepherds stopped grazing their animals there. The place began to fade from memory. There were even people who thought it was only piles of stones, common in the area, mounded up by those clearing the land for growing. Some of these piles, known locally as *majanos*, have been turned into bothies of sorts, shelters for the local shepherds (their integrity something of a mystery, given the lack of mortar), as well as way-posts and markers for the edges of game preserves. But Castillejo del Bonete served none of those functions. It can be hard to tell a *majano* from archaeologically valuable ruins that have devolved into mounds over time.

Save the shepherds, the occasional farmer and whoever it was that gave Castillejo del Bonete its name, everyone else assumed nothing had been there before. Moreover, some historians of antiquity considered us to be living in a Neolithic desert, and that our pre-history was an eternal night surrounded by an impenetrable darkness. Simply put, they could have learned plenty from the shepherds, if they had bothered to listen. While they kept on saying there was nothing there, the shepherds told their children and grandchildren what was there. For the rest of our elders, the settlement was "from the time of the Moors", because that was what had been written in the ancient chronicles.

A long while later, in 2000, when the local council commissioned a study and signed off on the initial excavations, we learned the true nature of the treasure our grandparents had dreamed about. Sometimes you don't look for what you suspect might be there, in case it turns out not to be. Better to leave it in the story realm and pass it down to your grandchildren so they can decide what to do about it, whether that be undertaking excavation work, or planting olive trees and allowing history to follow its natural course in the shape of fireside stories in winter or a yarn while getting a breath of air in summer.

But the archaeologists, with the support of a number of locals who knew the terrain, continued with their surveys. As the local population progressively moved to towns and cities, and archaeology became a way of preventing such a rural place from being forgotten altogether, hydrogeologists and archaeoastronomers also arrived, and we began to learn about what happened more than 4,000 years ago in this place about which we'd known nothing except that it was a good spot for stargazing. Today it has Starlight certification, but others knew that long ago.

Our ancestors from the Chalcolithic and Bronze Ages built a sun altar there that, seen from the sky, looks like a circular labyrinth. It all revolved around a cave that was created by the rain, scarce yet persistent. For centuries people placed their ancestors in the cave, along with valuable objects, and made some rudimentary paintings at a point where a ray of sunlight filters through. The galleries that housed the oldest burials can be reached only by crawling on all fours through dark, damp hollows. Here the sunlight touches only one spot, illuminating a cave painting (anthropomorphic, red) located by the feet of some human bones. The senses being deprived, you can hear the drip-drip of water only after recent rains, which produces a curious effect: it is possible to see the sun outside and continue to hear the patter of rain at the same time. There are even moments when the light catches the drops in such a way that they look like stars tumbling down inside the cave. In those conditions, while holding torches, the prehistoric people of La Mancha would crawl, dragging their dead to deposit them as deep into the belly of the cave as possible. One day, after 500 years, the cave was sealed and burials began to take place in tumuli built in the surrounding area and joined together by corridors. Thereafter, nobody went inside for another 4,000 years, when a local man enlisted by the archaeologists disappeared down a hole and came back a little while later with a handful of bones.

Though we do not know why the cave was sealed or the tumuli were built, these were not spontaneous decisions, nor were they exclusive to Castillejo del Bonete. Shortly before that time, across the Iberian peninsula, communal burial sites ceased to be used. Certain people started being buried apart, with belongings that indicated their status and usually included a bell-shaped urn. Young warrior chiefs tended to have their weapons and archer's armguards with them. The phrase "the richest in the cemetery" might not yet have existed, but its seed was there: inequality had just been born. People were no longer hidden away for burial, but rather located very visibly, laying claim to the territory.

The distribution of the prehistoric corridors and tumuli around the Castillejo del Bonete cave is not random. One of the corridors is oriented towards the peak from which the sun emerges on winter solstice. Rituals are believed to have been held there in anticipation of the arrival of what could have been their god, the sun, which also shone along an expressly carved, tapered corridor at sunset on summer solstice.

Doubtless in honour of their ancestors, during these rituals they drank alcohol and shared food, as in modern-day wakes. As offerings, they sometimes left food or personal items, before sealing the burial mounds. Most of the bones appear to have been moved from place to place, a custom that, according to recent evidence, was being practised even further back in prehistory. The ritual was probably very similar to the *famadihana* or "turning of the bones" of the Malagasy people in Madagascar. There they take out the bones of their dead, clean them, change their shroud and hold banquets and festivities in which they make the dead dance, before returning them to their graves. Honouring your ancestors thus becomes an excuse to bring the family together in an event that is organized years in advance to ensure that no one is missing. What for many would be desecration, for them is an act of familial love. Possibly this was also the case for the prehistoric inhabitants of Terrinches and other parts of the region from which they brought their deceased to rest in Castillejo del Bonete, judging by a funerary stele that includes both seashells and fossils left there by someone after a fifty-kilometre journey.

At some point in Castillejo del Bonete they also abandoned the ritual of turning the bones, and even stopped burying their dead there. The moment coincided, not by chance, with the end of a centuries-long drought.

To date, only four burial mounds with human remains have been found outside the natural cave. One of them has allowed us to see the face of a prehistoric man from La Mancha, whom the locals anachronistically named Luciano, in honour of their patron saint, the Virgin of Luciana, to whom they have prayed for rain on many an occasion. Luciano was a man who used a bow and at some point suffered a blow to one eye. He recovered from that, but died some time afterwards, aged 40 or 50, with back problems associated with age and the use of the bow.

He was not the only Bronze Age ancestor we had the chance to meet. Another surprise awaited us among the tombs of his neighbours: a secret hidden for 4,000 years inside a hill crowned by piles of stones that would soon become known around the world thanks to the largest ever study of prehistoric DNA.

A woman and a man side by side on the ground in the foetal position, as if the person who laid them to rest like that had hopes of them one day being reborn, their most precious objects alongside them: water vessels (in case they got thirsty?) and copper knives. He died before her. Doubtless wishing to have his status as a warrior reflected in the hereafter, he wore an archer's brace, while a handful of reeds and a rivet were located level with his waist. He also took a ridged bowl with him. Some time later, when she died, they opened up the tomb in order to reunite the couple. Their luggage for journeying to the afterlife included a cooking pan with an engraving tool and a small knife inside it. His shroud had a pair of ivory buttons. After placing them in spooning position, someone scattered soil over them and a fire was lit, into which a copper-tipped arrow seems to have been shot. More earth was scattered and the site was dressed with stones. And under the remains of the fire the couple remained curled up forever.

That intimate moment lasted until archaeologists burst in on them. They had been positioned facing east and the bones were undisturbed, though the collapse of the overarching barrow had left heavy stones resting directly on them, and there were signs of a fire having been lit at some point on subsequent layers of earth. Her buttons had fallen onto her breastbone, but unlike the disintegrated fabric of her garments they were like new. He was in his thirties when he died, she was somewhat younger. Their bones, the most recent in the dig site, are the only ones that appear in their original position. Nobody went back to turn them.

At first sight, there is nothing extraordinary about Tomb no. 4 in Castillejo del Bonete: a couple died and took to their grave certain shared objects. An isotopic analysis revealed that her diet had included marine proteins, but she was also found to be related to other local women; perhaps she had returned to her ancestral lands after a life of travel, or after living somewhere coastal like El Argar, where the ivory buttons, of African origin, may also have been acquired.

The pair were descended from two different migratory waves from the Middle East that took place thousands of years apart. Although she had always lived in Iberia, genetic traces were found in her mitochondrial DNA of people who, with their great exodus, brought agriculture and livestock farming from Anatolia 9,000 years ago, when the Peninsula was one of the few habitable places in this part of the world. Those first farmers settled throughout Europe, in valleys and on hillsides, always close to rivers, in search of fertile land which they soon deforested using slash-and-burn techniques to plant their cereals and legumes.

He had not eaten the same marine protein she had. His genes spoke of a place more distant even than El Argar. Although his lineage had existed in the same land as his lover's for several generations, his ancestry turned out to be another revelation.

The wheel. The domesticated horse. Brown eyes. Marijuana. Language. The lactase gene, which enables adults to tolerate lactose. The list may seem random, but all these things have a common denominator: the people who, according to various studies, seem to have disseminated them throughout much of Europe and parts of Asia. They were herders who lived on the steppes between the Black and Caspian Seas, in what is now Ukraine and southern Russia. Their culture seems to have emerged on the banks of the Volga, although some geneticists believe they descended from Armenians and Iranians. The Yamnaya lived on immense steppes, they rode horses (seemingly they were the first to domesticate them) and they were driven by thirst. Some of them may have returned to the land of their ancestors in the southern Caucasus around 5,000 years ago, but a large number of the Yamnaya set out on an immense journey that was to change European genetics forever.

Shortly after the emergence of Sumer and around the time that Egypt came into being as a unified kingdom, the descendants of the Sumerians began to expand across almost all of Europe, the Iberian peninsula included, bringing their own innovations and others they would have encountered along the way, as well as the steppe genotype. Although they were nomadic herders, some now settled beside rivers to try their hand at farming. In a very short time their burial mounds, which Lithuanian archaeologist and anthropologist Marija Gimbutas called *kurgans*, were present beyond this people's place of origin, and forests far and wide were once more being cleared.

By the beginning of twentieth century a search had been underway for 300 years for a possible mother tongue to the great

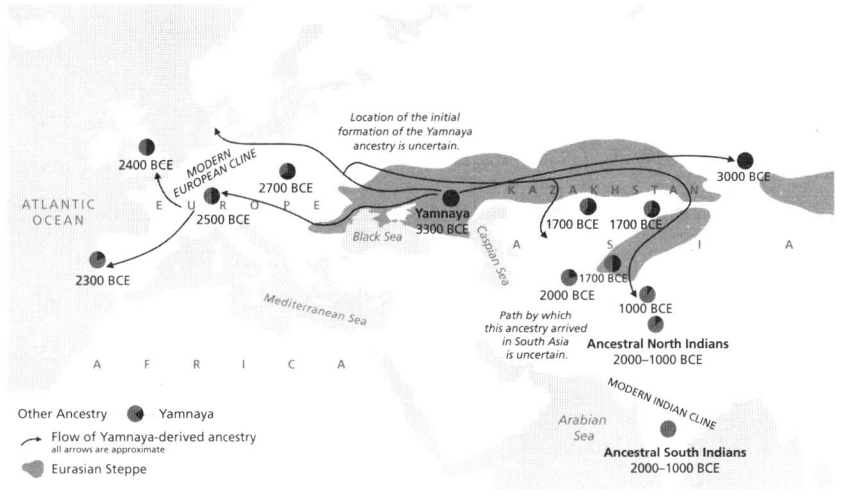

The Yamnaya journey

Indo-European group of languages. But when an archaeologist and a linguist then posited a hypothetical mother tongue for various European and Asian languages, the idea quickly assumed political dimensions. As thoughts were turning to the possible speakers of a proto-Indo-European tongue, the belief in an Aryan race—and its superiority—also arose. But "proto-Indo-European" was a way of referring to this hypothetical language, and not an ethnic group at all. Nevertheless, certain people were keen to give it the status of a race, and decided to make it a blue-eyed, fair-skinned one. When the Yamnaya people (who were not precisely blond or blue-eyed) arrived in westernmost Europe, the population was already eminently mestizo: *Homo sapiens* and *Homo neanderthalensis* had been interbreeding ever since their first encounter on the continent more than 40,000 years before, until, come the end of the Ice Age, they found themselves backed into the cold, arid corners of the Cantabrian coast and Black Sea. What is more, the Anatolian farmers—the girl's ancestors—had only begun their expansion across

Europe a few thousand years before. Finally, the Yamnaya arrived, having taken a similar route but considerably more swiftly, thanks to their horses and wheeled conveyances. Therein the supposed purity of the so-called Aryan race, which in reality referred to people who may have come from the Caucasus and Iran, who in turn came from somewhere in Africa, along with the rest of humankind. Not only did the Yamnaya not bring blue eyes with them, but they are believed to have replaced them with brown in some places. Fortunately, ancient DNA studies are scientifically demonstrating the absurdity of certain ideas that seem to arise when some people look in the mirror. Archaeologists say that the image provided by the couple's tomb is the closest thing we have to a photo of our true Iberian ancestors.

Those who bore the steppe genotype remained here, going on to replace almost half the Peninsula's population—the entire male population. Within 500 years they changed the genetic make-up of Iberia, and of much of Europe, forever. Those who have studied the remains of my ancestor (who, ultimately, turned up in an olive grove belonging to my great uncle) say that people nowadays born in the Iberian peninsula still have forty per cent Yamnaya DNA. When these conclusions became public, before the article was published, some researchers threw their hands up in horror. Archaeologists urged caution, given the way purported Proto-Indo-European-speaking steppe herders had been turned into a biologically superior race that fuelled Nazi ideology. Later, the Yamnaya again caused a stir when certain media outlets used terms like "invasion" and "extermination" in describing the wiping out of the Iberian male. Ill-advised talk, especially given the precedents, and not at all supported by the evidence.

Without evidence, it is rash to assume that such situations can only occur through orchestrated extermination. Similarly, there is a tendency to blame Sapiens for the extermination of other human populations, Neanderthals included, despite the fact that they coexisted, as we have seen, for something in the region of 10,000 years and that, when they first came together in Europe, the Neanderthal population was already very much reduced. We have hardly any fossil or skeletal remains of either the first European Sapiens or of the Yamnaya, just a few traces and a smattering of objects. We cannot know their thoughts, what decisions they made or what their intentions were. Nor are there archives or graphic evidence which might tell us about how they were eliminated as a species, so we cannot speak of genocide based solely on our own biases; genocide is systematic, premeditated and has the clear objective of wiping out a specific group. Evidence would have to be found, such as, for example, mass graves, to state with any degree of rigour that there was an invasion or extermination. Feeling nostalgic for the killer-ape hypothesis is not enough.

Homo sapiens and *Homo neanderthalensis* coexisted with other species that also became extinct, and we have recently learned of the great hardships endured by our ancestors at the end of last Ice Age. Shortly before leaving Africa, the *Homo sapiens* population had been reduced to around 1,000 individuals by another ice age; this was the same population who were once again pushed to the limit 70,000 years ago after the eruption of the Indonesian volcano Toba, when we were still sharing the world with *Homo neanderthalensis*. And at both those moments in history, something very similar happened: the tiny group of surviving *Homo sapiens* and the Yamnaya who were among those to have descended from them may have been better prepared than respective contemporary groups to pull through. When the latter arrived on the Iberian peninsula, barely twenty to thirty per cent of the local hunter-gatherer genes remained among the population. Their genetic heritage, like that of *Homo neanderthalensis*, had been diluted thousands of years earlier by the Anatolian incomers.

It is undeniable that the arrival of *Homo sapiens* coincided with the extinction of other animal species, and of humans as well. But our ancestors, as we have already seen with *Homo erectus*, did not travel alone, rather setting off on journeys in concert with other species. Feral cats, for example, are a serious risk to species endemic to some islands, but does that mean that they are mass killers who plotted the extinction of the La Gomera quail? Some people see a wall and immediately think of wars and invasions, even if it was built to protect the dead, just as others see the grave goods of a warrior and assume, before the analyses show otherwise, that they belonged to a man rather than a woman. We are neither Rousseau's saints nor Hobbes's demons (I refuse to use wolves in this context). And at the same time, we're both. The ability to choose between a good and an evil that we too have created, and the ability to accept that we have as many upsides as downsides, makes us human on an individual level, and neither better nor worse as a species. Between a naïve and a suspicious view of humankind, I choose the trusting option until scientific evidence proves otherwise.

But back to steppe herders. All scientists know for now is that the period over which the replacement of the men on the Iberian peninsula took place is too long for it to be considered genocide. They have not yet determined whether the substitution was due to the Yamnaya bringing a disease like the bubonic plague, or whether Iberian males failed to survive a concurrent change in climate, or whether it was something as simple as the women being attracted to novelty. Therefore, we cannot categorically affirm or deny this either. But to simply turn women into victims of the newcomers is automatically to deny them the possibility of choice at a time when, according to recent studies, they too hunted, fought wars and travelled great distances. Evidence of this has been found in Ukraine, Peru and Sweden. But we don't even have to go far to discover that they also held power. As I write, archaeologists have found what could well be the "lady of the water" at the Bocapucheros site (Almagro). This is a tumular funerary monument complete with corridors above a cave in the style of Castillejo del Bonete, similarly located in an elevated area with a specific astrological orientation, in this case towards the Southern Cross constellation. In one of the burial chambers the remains of a woman have been found lying in the foetal position next to a water vessel. The type of burial reveals that, in a hierarchical society such as theirs, she may have wielded power. Her power, unsurprisingly, lay in her ability to command and manage water. And she was not the only woman buried in the chambers of that sacred place.

Beyond an ideal around what might have constituted good looks at that time, we know that the Yamnaya and their descendants had a mastery of technologies previously unknown in the area, as well as the gene that allowed lactose to be tolerated. Maybe there was no invasion, kidnap or extermination. Maybe the local women simply preferred the idea of the new arrivals as fathers to their children; we can't claim that they were the mountain climbers of prehistoric Tinder minus the profile pictures, because not only did those not exist, they also didn't need them; showing off how to use a wheel or ride a horse would have been more than sufficient. With their wheeled carts, they had attained a measure of independence from water sources, as they could now carry with them all they needed—food and materials for shelter included. They also arrived with surely the greatest possible turn-on for a proto-Manchegan woman in the

Bronze Age: they knew how to make cheese. Though this might sound like my devotion to that ancestral foodstuff talking (which it also is), it has been shown that lactase is not only a selective advantage, but allows for a greater absorption of water from milk in arid places. It is now the most widely dispersed gene in the whole continent. Meaning we can conclude that those people knew how to solidify milk as a way of making it last, as well as how to transport it and reduce the lactose content until it became a delicacy that we Manchegans simply could not now live without. This helped both our survival in a dry climate as well as their own. Invader, refugee—these could be one and the same, depending on who is doing the telling. And the Yamnaya had good reason to be climate refugees, just as much as those people who at the same time were setting up camp on the banks of the Nile and went on to be the founders of Egypt.

Whatever the case, the jury is still out on what it was that brought the Yamnaya to these shores, but we do know that the move had taken around a thousand years and that they left behind lands that were becoming increasingly arid and harsh. If this was the case, I can't help wondering why they didn't stop on the banks of the Uzboy, in modern-day Turkmenistan, a river that was much closer and, being out in the middle of the desert, had attracted endless numbers of people in need of water, giving rise to a new civilization practically out of nothing, at the exact moment when the Yamnaya started to leave their lands. It debouched into the Caspian, one of that people's seas, after crossing the Karakum desert. Perhaps the answer is simply that some of them had made it to Iberia previously, and that that civilization, like the river that sprang up in the middle of a desert, began to peter out not long after it had flourished.

✽

The average life expectancy of people in La Mancha a little over 4,000 years ago was like that of modern-day rock stars: twenty-six. Although some lived into their forties, fifties and even sixties, this would have been in the constant grip of arthritis. Their lives centred around the raising of livestock, to judge by the remnants of cheese dishes, strainers and looms in their homes, as well as the sheep bones among the offerings in Castillejo del Bonete. Sheep, goats and cattle: not only would they obtain milk, cheese, wool and meat from these, but also the power to pull a plough. These doubtless were the first Manchegan herders, but rainfed and cereal-based agriculture supplemented their subsistence, especially in the fallow, permanent pastureland immediately around their villages. They grew wheat and barley, as well as legumes such as peas, lentils and haricot beans. These they sometimes planted at once, having learned not to rely on the rain; if one crop failed, another might make up for it. Everywhere around was meadowland, mixed with small, dense Mediterranean forests of cork oaks, holm oaks and common oaks, but also bushes such as strawberry, rock rose and lemon verbena.

Life was not easy in such harsh climes. Prolonged droughts were already frequent and cyclical, because the butterfly that flaps its wings and causes a hurricane in another part of the world has its equal and opposite in the dance between the atmosphere here and the Pacific Ocean. When the Pacific's atmospheric pressure is high, it triggers a process that ends up causing droughts both in India and in the mountains of Ethiopia. Which is just what happened. Four thousand years before these people settled in La Mancha, the Gulf Stream shifted and the climate suddenly cooled, increasing both the number of icebergs and the amount of fresh water in the North Atlantic. The same phenomenon repeated itself several times over. But added to that, the oscillation of the Earth's orbit also modified the solar radiation received by our planet from 2450 BC onwards, changing the rainfall pattern until 1850 BC. The people had to face a new mega-drought that scientists have called the "4.2-kiloyear BP aridification event". It was the most prolonged and intense of that period and may have been the worst in millennia. It had devastating effects on much of the world over several centuries. Its onset coincided with the dispersal of the Yamnaya and the origin of a new culture, that of the Motillas. Particularly badly hit were North Africa, the Middle East, the Arabian Peninsula, the Red Sea, the Indian subcontinent and parts of North America. However, some climatologists believe its effects were fully global.

But it was no mere continuation of what had been naturally occurring for millennia; in addition to the long-standing astronomical causes, agriculture and livestock farming had spread, laying waste to the land, causing deforestation and bringing even more drought. Entire forests disappeared and several tree species were brought to the brink of extinction as humans cleared space for pasture and larger expanses of arable land. Devoid of trees, the soil became increasingly dry. It stopped raining. Life became intolerable and civilizations around the world collapsed. The life stories of the couple buried in Tomb no. 4 are framed within a period of climatic stress that made for no rain anywhere in the world for years, decades or centuries, depending on who is keeping count. The Yamnaya homeland was no exception. Surrounded by vast steppes that they crossed on horseback, and a long way from any water, thirst forced them into nomadism, and over several generations and with many stops along the way, they ranged ever further afield. Nothing unusual: the search for water has made both nomads and settlers of all of us.

❖

Thirst was now affecting large parts of the world. In the city of Ur and all around it the winter rains stopped coming. Around 4,300 years ago, the Akkadians—Semitic, drought-fleeing nomads with a leader called Sargon the Great at their head—went off in search of fertile lands in other parts of Mesopotamia. They conquered Sumer and, after uniting various peoples, founded the Akkadian empire—history's first ever. Its cities were even visited by nomads driven from the Arabian desert, whom they called Amorites, and who came to quench their thirst and that of their livestock. Nomads from the Zagros mountains also arrived both in Akkad and Sumer during this time. But the Akkadian Empire was wracked by a drought that forced northern herders south, and in both Akkad and Sumer the Amorites, seen as belligerent water-thieves, were less than popular. A menace.

Their intentions however may have been misrepresented; perhaps all these herders wanted was to bring their animals to water. Perhaps the Akkadians saw themselves in these newcomers and didn't like what the mirror showed. To protect themselves from invasion, the people of Ur built a wall to keep out the Amorites. The plan was a terrific failure. The population tripled and the city was not set up for such demographic pressure. Ur collapsed due to a combination of a sixty-year drought, dust storms and social unrest, and was eventually taken over by the Elamites. The wall was toppled. First the Guti (tribes from the Zagros mountains) and then the Amorites took over the empire, impelled by the very thing that had sent Sargon the Great into those lands to begin with: thirst.

Dust storms forced the population of Akkad to abandon their city. It would not be resettled for another 300 years. History's first empire, barely 100 years old, had just fallen. In the city of Ur, someone wrote a poem called "The Curse of Akkad": "On your steppes/ where succulent plants once grew/ tears shall now spring forth…"

For a long time, various explanations were considered for the fall of Akkad and the sudden abandonment of its main city. Both the fossilized dust and this poem have kept the answer hidden for thousands of years. Another line from the latter said: "Thick clouds did not bring rain." Recent studies of the fossilized dust in stalagmites and the balance between worms and soil in a site in present-day northern Syria have found "something that choked the land with dust for decades", and which drove the people out of the city in 2230 BC. The situation affected Mesopotamia to the same extent as it did the Nile, the Aegean and the Mediterranean. It was the same extreme drought that devastated the Middle East and much of the world besides; together with demographic pressure, it brought several cities to ruin. When the Nile rose, it did so only relatively moderately, and given that post-diluvial silt was the basis for life throughout the rest of the year, this proved catastrophic. Thirst and hunger took hold, unleashing violence in the streets. A testimony remains on the tomb of King Ankhtifi: "The entire population has become locusts in search of food." It was not long before the Old Kingdom of Egypt was in crisis, and effectively disintegrated. After a series of droughts, the Harappan civilization also fell and the Liangzhu in China were in desperate straits. Just a few centuries later, 3,800 years ago, the inhabitants of Caral carved friezes depicting frog-like people with empty stomachs that seem to tell the same story as the pharaoh's tomb. It may have been a way of invoking rain, as they even marked on the stone the direction that the rain should follow, in the style of the water channels and basins of La Mancha. But no rain came, everything was filled with sand and the people of Caral had to abandon their lands.

More and more drought-related collapses have come to light, like that of Rapa Nui on Easter Island, of the Hittite and Mayan cultures and of the Tang dynasty in China. Other factors are always at play, often caused by thirst itself and by the decisions and actions of rulers, but it is no coincidence that drought has been called humanity's worst enemy. For Jared Diamond, changes in climate and mismanagement of the resources on which these civilizations depended, especially water, are among the main reasons for these collapses. He has called this combination "unpremeditated ecological suicide". Some collapses and crises have recently been studied with a climatic frame of reference and the same conclusion is always reached: there were other factors, of course, but they followed a series of prolonged droughts and many occurred during the 4.2 kiloyear aridification event. Not even the kings who thought themselves gods in some places, and messengers to the gods in others, turned out to be impervious when the dust storms came. The Chinese and Sioux alike have a warning proverb that runs: "The frog does not drink from the pool in which it lives." It was precisely this that the Garamantes people—who settled in the Sahara a number of centuries later—failed to understand. All surface water had disappeared there, but they discovered an immense aquifer underground. By building *qanats*, they managed to bring water to the desert and had use of it for centuries. But, made too bold by the favourable climate, they tapped the underground reserves for too long and in an unsustainable way. They ran the aquifer dry, hastening their own end around 2,400 years ago.

Here I want to digress to share an idea from the anthropologist James Scott. According to him, the word "collapse" should be selectively deployed, because it often means confusing kingdoms and civilizations; sometimes only kingdoms fell while the civilization continued. The Old Kingdom of Egypt collapsed, but Egyptian civilization continued; the city of Ur fell, but was reborn in time; and the empire of Akkad too, but life went on in Mesopotamia. Sometimes people left the cities only while drought or dust took hold, but the former inhabitants or their descendants returned when the conditions relented. Often, too, what we call "collapse" is nothing more than a pause or the fall of a ruler with a god complex who neglected earthly matters and failed to act in time.

But things did not turn out so badly in La Mancha, at least in the centuries during which that devastating drought lasted. The prehistoric Manchegans came up with an idea that allowed them to stay put and establish their own culture, coeval with that of neighbouring El Argar and with the hydraulic civilizations on the floodplains of the few rivers still flowing at that time.

While thirst unleashed catastrophe in Mesopotamia, Egypt, the Indus Valley and Caral, in a region that was about to move from the Chalcolithic to the Bronze Age, someone devised a structure that made it possible to extract groundwater, and it was then reproduced in several nearby places. Thus, together with those in Motilla del Azuer, Motilla del Acequión and at least forty others, the first wells emerged on the Iberian peninsula, forming part of what prehistorians consider the first hydraulic society. This was nearly two millennia before the Garamante civilization. Although drought conditions remained the norm, this was interspersed with times of rain, and even times when ditches had to be built around the Motillas to contain the water.

These curious constructions, apparently defensive, served to channel water, protect the wells, store grain and, in some cases, they were used as burial sites. They were interconnected, often over short distances, to keep various villages from running out of water. That advance in proto-La Mancha was doubtless a draw to thirsty peoples elsewhere.

We do not even know whether the invention was the work of the indigenous people or the Yamnaya, who hove into view seemingly at this very moment. There were *shadufs* in Mesopotamia by this time, which enabled water to be raised from a well. In Castilian Spanish we call these crankshafts *cigoñals* for the way they resemble a stork—*cigueña*—dipping its head in the water. Julio Caro Baroja claims that *cigoñals* were introduced by the Arabs several millennia later, but it would not be unreasonable to think that those who introduced the wheel, which was already in use in Mesopotamia, and also the horse as a draught animal, also bequeathed us a structure that required both. Moreover, at the time that *shadufs* had spread to places such as Egypt and India, people might have been using similar techniques with the help of ropes, wheels and draught animals. The *magrod*, for example, is still around in parts of Morocco today. The *sakia*, too, which when operated by animal traction is known as an *aceña*. This is the word used in *Don Quixote*, as well as to this day in some Spanish places.

The Motillas tend to be circular and are often arranged around a well in a central courtyard. Generally found on floodplains, in areas of low salinity, near rivers and in spots where the water table is near enough to the surface to be reachable with the tools and techniques of 4,000 years ago, they were established in places that were not constantly dry. It isn't known whether their founders ever saw for themselves surface pools or puddles before these all disappeared, or whether these people knew the technique employed by others at the time and that the engineer Marcus Vitruvius extolled almost three millennia later as a way of working out where to build your well: lie down with your cheek on the ground at sunrise and wait for mist to appear. It was roughly around then that human beings are thought to have discovered something that the Namibian Desert beetle also knows: where mist is, water can also be found. Or they may simply have been guided by the presence of certain plants that, as our ancestors may have known since time immemorial, point to the presence of groundwater. We will never know which piece of ancient knowledge was key in their settling some places and not others, but for the first herders of the Ruidera lagoons, it would have been the presence of an oasis.

Motilla sites are particularly concentrated in areas of natural surface drainage around Aquifer 23 like the Tablas de Daimiel national park, the Ojos del Guadiana springs, the Ruidera lagoons and a number of rivers besides. These constructions are particularly concentrated in the area around present-day Ciudad Real and also in Albacete, and to a lesser degree in Toledo and Cuenca, although in those cases their presence is quite limited. The main clusters are over the top of five groundwater aquifers in the Guadiana and Júcar river basins. Eight of the forty-five uncovered to date have been found exclusively around Tablas de Daimiel.

They were more than just wells: people were buried in the environs of the water, often with urns and other grave goods. They shared the area with the *morras* (high-altitude settlements), silos and ceremonial centres. Of the latter, until the recent discovery of Bocapucheros, there was evidence of only one, familiar to us by now—Castillejo del Bonete—which was situated between two Motillas, connecting the sun with the underworld, life with death.

This network of watering holes functioned for nigh on 1,000 years. It managed to hold out while civilizations elsewhere collapsed—civilizations that could already count on large cities and irrigation channels and in which writing had also been developed. Wittfogel suggests that China's superior resistance compared to other civilizations was attributable to the fact that it was based on hydraulic structures such as dams and canals, making it less dependent on the rain or indeed the consequent rising of the rivers. The case of China is, in this regard, comparable with the proto-Manchegan culture that made use of aquifers, which gave a steadier supply than the rain.

In our contemporary world, these constructions would not have lasted; at the time of writing, the well at Motilla del Azuer, which provided water during a drought that lasted many hundreds of years, is now dry, in spite of some rains which, though late, have brought reservoir levels up. The aquifer underneath Tablas de Daimiel has been so overexploited since the 1980s that there is now water in only two of the three channels, and this only because of emergency injections from other rivers. The river skirting the Motilla is now completely dry.

As discussed, the collapse of the Akkadian Empire in Mesopotamia, the abandonment of the city of Caral and the end of the civilization that bore its name in what is now Peru—the first in the Americas—have all been linked to that mega-drought. The collapse of the Harappan civilization in the Indus Valley and that of the Konar Sandhal in modern-day Iran have also been associated with the 4.2 kiloyear climatic event. In dry Iberia, during the same period, the aridity had lasted some 600 years, affecting the vegetation and the landscape as well as the population. There was a sharp demographic drop in the southwest; agricultural and livestock activities in Doñana were abandoned and the number of people living in the marshlands decreased. Meanwhile population numbers increased in the southeast and stabilized in the north, which coincides with both the beginning of the drought and the arrival of the Yamnaya in a place where Chalcolithic indigenous people had already settled.

Imagining a network of Motillas connected to ensure that several settlements would not go without water brings with it a tempting thought: that my ancestors unified to tackle thirst in a harmonious way, like the black worms in California that press themselves into a ball in their thousands when drought worsens, in order to retain what little moisture they have accumulated. Yet nothing could be further from the truth; it is believed that there was a "water lord" in play here, whose dwelling was at the high point of the highest settlement. So inequality already seems to have been a feature of life on the Peninsula, depending as it did on control of water access and the stockpiling of grains that had adapted to an increasingly dry climate and were crucial to survival—and crucial to the hierarchy as well, we might suppose. A halberd has been found at Motilla del Retamar (Argamasilla de Alba), an object whose only purpose can be violent killing, and which shows that these places were no hippie communes. Even so, it is the only example turned up in dozens of Motillas, where it is also possible that a collective ambition existed, since the wells were not personal, independent or isolated, but rather formed a network and had acquired a special, even sacred status, based on shared beliefs that we will surely never fully understand and that prompted people to travel to Castillejo del Bonete both to connect with their ancestors and the stars and to bury and be reunited with their dead.

Approximately 5,000 years ago, descendants of the hunter-gatherers from the Caucasus moved into Armenia, which was their ancestors' land, while others migrated to the Balkans and Greece. In the following centuries they continued to spread towards both Iberia and India. Some of the words they spoke seem to have remained in our languages to this day. Of these, the ones relating to emotion tend to be negative, like fear, terror and hate. But those that are land-related are very often linked to water.

There is a peculiar coincidence that I find pleasing to think about: *yama* is Russian and Ukrainian for a "hollow" or "pit", and the Yamnaya were renowned for the way they buried their dead, firstly by digging deep holes and then by laying the body on its back with legs bent, followed by the scattering of soil. Over these they built their *kurgans*, mounds that in time came to look natural; for Gimbutas, so influential in transferring the steppe hypothesis from linguistics to archaeology, this was their sole archaeological trace. She frequently put forward the idea that burial rites might have led people to locate headwaters accidentally. Without claiming that this is what happened, or indeed that archaeology has suggested as much, we can still notice that their sepulchres did resemble wells, and the descendants of the Yamnaya arrived just before the first ones were built, in drought conditions. Plus, prehistory is full of such eureka moments which help to explain how fire, plants, animals, bread or beer were discovered, and indeed mastered, by complete chance.

Terrinches's steppe descendants were here for several generations. As far as we know they performed Castillejo del Bonete's second-to-last ever burial. And though not perhaps in this exact place or in that exact moment, they were surely always looking for water even when that wasn't their express intention. The subconscious sometimes has its own reasons, excuses and gestures, and every time humankind followed some animal's migration, or went in hunt of fertile land, this at root was their search for water. Or it might have been someone interpreting animal behaviour, or the presence of a certain plant; the language encrypted in the land. Sometimes—almost always—survival happens unexpectedly and unexplainably.

Some hydraulic societies coeval with the Motillas considered rivers sacred and, like the Manchegans of the Bronze Age, buried their dead near water. In later times, Muslims, whose religion arose in a desert, buried their dead near rivers, which they too consider sacred; they also still have a practice of placing a bird bath near the grave, inviting the river birds to drink and thus carry the souls of the deceased in their bellies all the way up to Allah. Evidence of this ritual, the *hadith* of the green birds, has been found at a dig site in Villanueva de la Fuente, beside the Villanueva River.

The Motillas culture held out, making use of those wells for nearly a thousand years. But all proverbs come into being for a reason, and I think of the one we have here: "To every drought, a downpour." At some point, so it was, the rain began to fall. And when it did, in around 1900 BC, small dykes had to be built alongside the Motillas to hold back the water. But these eventually fell into disuse. Some even flooded when the climate turned topsy turvy and the rain really began to fall; they were not made to hold so much water. Thought had gone into extracting groundwater when there wasn't any rain, but not how to store unhoped-for amounts of it. Not long afterwards, in 1400 BC, the crisis of the Manchegan Bronze Age began, and the Motillas were abandoned. Some opted for livestock farming and moved to new or established settlements in higher areas.

Wheat, adapting to the extreme aridity, continued to grow in the Motillas, as evidenced by carbon remains of seeds at Motilla del Retamar. The presence of sheep bones among the grave goods there has raised questions about the origins of Manchegan sheep farming. It is a curious detail that their descendants, Terrinches's most recent herders, took their sheep there, presumably themselves wondering if any water was to be found under the stones.

When Castillejo del Bonete was discovered, the first hypothesis put forward by Luis Benítez de Lugo, the archaeologist in charge of the excavations of the sun altar and of several other Motillas, was along these lines: there must have been groundwater there, since that was the case in other nearby Motillas. But both he and Miguel Mejías, a hydrogeologist specializing in Manchegan watercourses, went on to disprove the hypothesis; if the cavity under the galleries had been vertical, the water would have been about forty-six metres down. And they would have had a hard time, with the means available, digging as deep as a fifteen-storey building. The wells found at other Motillas are no deeper than twenty metres and are also located on flood plains, very close to the water. Those were further distinguished by the fact the earth below them—low-permeability clays and limestone—was far easier to dig through. If someone had built a Motilla with a well here, they would have done so at the *foot* of Castillejo, in the exact spot where, thousands of years later, my grandfather extracted water for his final crops of tomatoes.

Castillejo del Bonete was not at a suitable height to be considered a fortification, but neither was it the best place for getting at the groundwater. It was on a hilltop, and had been built with white limestone, which was not exactly intended to make it go unnoticed. What was it doing there? It was very likely intended to attract attention. To use the dead as a banner and send a tacit message to people from southern and eastern Spain that more or less said: our dead are here, so this land is ours.

✿

Thousands of years later, my parents and neighbours bought water tanks and placed them out on their respective terraces. I often went out and lifted up the lid on ours, wanting to see the Styrofoam ball floating on the surface. Partly out of a desire to get water for myself, but more to see that we had enough for the coming days. In a way those tanks were a memory of the Motillas. While writing this book, I have been constantly asking my family and former neighbours about the things we and our forebears went through; about what it is that thirst has made of us. Although, proceeding chronologically, I didn't begin the book with this story, it was really the seed for it. The pieces are all now falling into place, thanks to those conversations. I said before that my grandfather used to tap cave water to irrigate his crops, but I still didn't know what role he himself had played in a place known to me as the Minao, in the Huerta Soriano, there at the foot of Castillejo del Bonete.

At first, the Minao was a crack in the ground. It seems that my great-great-grandfather Norberto, of whom I previously knew nothing, and whose name we discovered only because of a computer glitch, started widening it in his own search for water. The result was the cave that I came to know. Or, not exactly. His son-in-law, my great-grandfather Pedro, along with other relations who lived in the area, went on digging deeper as the soil progressively dried out. They extracted a small amount of water.

But later on the Minao again dried out. It was then that my grandfather (also Norberto) and his cousins started to dig, eventually making it as far as the spring that our ancestors had not been able to reach. But it was no easy task. At first all they got out was silt. It fell to my mother and aunts to carry this away in buckets their father had fashioned from old tin cans and bits of metal. But soon fresh water started appearing and my grandfather built a basin, leaving out an empty chopped tomato tin for anyone in need of a drink. Only the children were exempt from having to crawl inside there. A retaining pool was built next to the Minao, but when water was scarce you had to slither deep into the cave with a tin, though just as much of it would spill as you carried it out precariously.

As a prompt I have a diary that I kept between the ages of nine and nineteen. My main concern in the early entries was whether it was raining, snowing or cold, and how this would be affecting my parents while they were out selling dried fruits and sweets at one of the village markets. A repeated source of joy in these entries are the trips to the vegetable beds with my grandfather. It was so incredible, I told my diary, going with him, getting to explore the caves, or helping to clear one of the tracks so that cars could get up there. He, who went from donkey to wheelbarrow and from wheelbarrow to moped and scooter, and who never had a car. Why did he take me cave exploring? To do what he always did up there, surely: gather the thing that over the years had forced so much of his family to leave: water. He always wore Wellington boots. Only he would know the reason, in a place where the earth was dry. Perhaps it was his way of invoking the rain.

Not long ago, my brother rescued a memory from oblivion. He remembered going with my grandfather to the vegetable beds one day, then collecting the hoes from the farmhouse where all the tools were kept and heading down to the lower plot, where my grandfather grew watermelons. In a place that was very specific for my grandfather but not so my brother, who was barely five or six, they began hoeing. Deeper and deeper they went, in a search that my brother didn't understand at the time. And then suddenly, wet mud began bubbling up. Soon it became clear water, gushing out. It is the kind of image that one keeps forever in a box, a box that will suddenly burst open one day, without our understanding the context or anything more about it. My grandfather set to digging because, just like my brother, he had a memory of being thirsty in his childhood and knew that once, long before, in that same place, he had seen water.

There is a film called *The Water Diviner* that is about searching, about grief and loss, about war. About almost everything except water. But Connor, the character played by Russell Crowe, is based on a real-life person called Joshua O'Connor, a farmer and *zahorí* (Arabic for "water mage") from Monbulk in Victoria, Australia. The writer Andrew Anastasios had been working on a history project, and came across O'Connor's story. Eighteen months after the end of the First World War, O'Connor had gone to Gallipoli in search of his sons, and succeeded in bringing one of them back.

Information regarding the true story was almost non-existent, so Anastasios decided to base a novel on the very little he had found, and that was later turned into a film. In one of the early scenes, Connor starts digging in the soil, and keeps on going down and down, creating a shaft with supporting beams as he goes. First mud starts bubbling up, then water. The well turns out to be a small lake in the middle of the desert. Connor is described by his wife—and this we guess to be creative licence—as a man capable of finding water in the middle of the desert but not of finding his own sons. In another moment, he talks about the fact that he comes from a country where it rains barely three or four times a year, meaning the search for water is an essential part of life. As to his method for locating it, he says: "You have to feel it." This is precisely the *zahorí* technique, or water divining, which is being able to feel the presence of water. A technique that had been deployed 5,000 years ago in Mesopotamia and Egypt with the use of a hazel branch, as attested in a cave painting in Tin-Aboteka, Algeria, although the first written mention of it appears in 1568 in *The Life and Times of St Teresa*. On seeing a *zahorí* at work, she could find no way of explaining it and assumed it must be a miracle. To the Church, however, it seemed a superstitious, Satanic practice, and the *zahorí* were cast as witches. Nowadays water divining is considered a pseudoscience based on pure chance. But *zahorí* still exist in rural places today, as do enthusiasts who claim to "have the touch", purportedly passed down through families for generations. For example, in certain villages in Aragón like Sos del Rey Católico, a belief was held until recently that any child born on Christmas night would have the ability to locate water; their left leg would tremble, or mosquitoes would follow them around. And then there is the story told by Vázquez Figueroa, who grew up in the middle of the Sahara. There he met one Manolo, who would go on to be his teacher. The man had a kind of gift, said the writer, which he called "smelling the water".

Like Connor and Manolo, my grandfather was capable of finding water in the desert. He could feel it and smell it, though I will never know exactly how. Maybe the watermelons he planted there, and the way in which they grew, sent him a message that only he could read. Who knows if he too would lie flat on the ground to wait for mist to appear at dawn, or if his leg trembled, if mosquitoes followed him or if he had some kind of tool to help in his search? I recently asked my mother if she had ever met a *zahorí* in the village, someone with the ability to locate water using a pendulum or Y-shaped olive branch. Neither of us considered the possibility that we'd had such a master in our own family. I didn't know, when I set out to write this book, that the water my grandfather extracted from the Minao was part of the same seam that would be detected years later by exploratory drilling near Castillejo del Bonete.

In Ryszard Kapuściński's *Imperium*, he talks about an old man he met in Turkmenia (modern-day Turkmenistan) who knew about wells both dry and abundant, about deserts and oases, about desperation and joy. Of those like that man capable of surviving in the desert, the Polish journalist said: "They know where the wells are, which means that they know the secret of survival and salvation. Their knowledge, devoid of scholasticism and doctrinairism, is great, because it serves life." For some time now I have been wondering about the source of this wisdom, so useful in life and to me so elusive.

The Minao is dry nowadays. My grandmother Francesca has cast the tale a generation further back, claiming that the first to start digging there was the father of my great-great-grandfather Norberto, whose name she didn't know (although the church records have him as Rogelio, the only mule driver in a line of day labourers). How many generations of my family have been driven by thirst, I do not know, nor how long we have been digging the soil in search of water.

That search was nothing new for them, just as it was not for the people who inhabited this land thousands of years ago or those who arrived earlier still, from even drier places. From the time of Lucy, the grandmother of humankind, to my grandmother, who prays to the saints for rain, there lies a whole series of milestones that help us understand the thirst we still experience today.

II
MANAGING THE RAINS

It rained when I was thirsty. That was why I dreamed of rain: because I was dying of thirst. Then it rained and I had something to drink, because the rain usually responded [to my request].

ǁKABO, SAN SHAMAN QUOTED IN *LA NIÑA QUE CREÓ LAS ESTRELLAS* (THE GIRL WHO MADE THE STARS) BY JOSÉ MANUEL DE PRADA-SAMPER

The taxi driver who takes me to Ayerbe asks if I mind if he turns on the radio and I say no. A trade unionist is saying that the drought, as well as creating problems for cereal planting, also endangers next year's harvest.

"What do you think?" says the man. "No one goes on pilgrimage to ask the saint for rain these days, so we don't get any rain, right?"

"Do you believe in God?"

"No, but there's no harm in trying."

ARTURO SAN AGUSTÍN,
PLUMA DE BUITRE (VULTURE FEATHER)

6

Heavenly horns

"From the first human handprint on a cave wall, we're part of something continuous."

BASIL BROWN IN *THE DIG*

"Water is everything," says Ogotommelli, a wise man of the Dogon people, who live in Mali. "The earth comes from the water. Light comes from water. And blood."

RYSZARD KAPUŚCIŃSKI,
THE SHADOW OF THE SUN

Ogotemmêli shot at a porcupine and the rifle exploded in his face. Left permanently blind, the hunter recalled the resounding message a fortune teller had once given him: "The work of death attracts death." At the time the rifle exploded, Ogotemmêli had lost sixteen children, so he took the accident as a final warning. He shut himself away in his own world and gained a reputation as a wise man in the Bandiagara Escarpment of Mali and beyond. After many attempts by the French anthropologist Marcel Griaule to get him to speak, the Dogon sage decided to share the beliefs of his people. The first thing he said was:

"Greetings to all who are thirsty."

Anthropologists had spent the previous fifteen years searching those lands for the stories of the Dogon, whose name for themselves is *habe*, or "nonbelievers". For centuries they have lived in an area of inaccessible cliffs along a 200-kilometre fault line. Their provenance is unclear, as are the dates and reasons for their arrival in the area. So mysterious are their origins that some even spoke of extra-terrestrial ancestors. They, on the other hand, believe themselves descended from Sirius, the star that, like the mythological heron Bennu, announced the flooding of the Nile to the Egyptians. And indeed their line has been traced back to Egyptians who fled to this area in the fourteenth or fifteenth century. Fleeing what? According to some, from being converted to Islam, a religion that, incidentally, had itself expanded largely on the back of drought, as a recent study of stalagmite growth has shown; according to others, from a drought, but only as the latest in a series of migrations through different lands over the centuries. The proximity of the Niger River, and of a stream that refills whenever it rains, would have recommended the place just a few centuries ago. Griaule himself dated their existence back 5,000 years, and considered the Dogon he met the true guardians of Egyptian wisdom. His daughter, Geneviève Calame-Griaule, studied their language and her thesis ended up in the hands of the philologist Jaime Martín. There were words and structures in it that he found familiar. He embarked on a comparative study that lasted more than a decade and led him to look at thousands of words in Dogon and Basque. Finding similarities in close to seventy per cent of the lexicon analysed, as well as structural resemblances, he paired the two languages. Martín is convinced that the Basque language derives from Dogon and that the relationship is due to sub-Saharan desertification that would have driven some Dogon as far as the Iberian peninsula long, long ago.

Whatever the truth of that, Ogotemmêli's testimony spoke to the age-old link between Dogon stories and thirst. Over the course of thirty-three days, Ogotemmêli told Griaule that the stars were balls of mud flung into space by their god, Amma, a potter who baked Sun and Moon in his kiln and one day rolled out some mud, carpet-like, to create Earth, a feminine body that soon awakened the god's desire. As he approached a termite mound, a representation of Earth's clitoris, she stood up and the god discovered that she was in fact a being just like him. He went away in disappointment. But he didn't stay away. Water appeared in the world as divine semen, and Earth was left naked and mute. From this primordial violation, the jackal was born. Amma created two Nummos, water creatures that were half human and half snake. The word translates as "to make water". The conjoined pair of Nummos, amphibian, hermaphrodite water spirits, would be present wherever water was found.

"The life force of the Earth is water," said Ogotemmêli. "God moulded the Earth with water. Blood is made out of water. Even in a stone there is this force, for there is moisture in everything."

The Nummos sent down language from the sky through the threads that made up Mother Earth's dress. They brought forth the first word, and the second and the third, conferring order on the chaos unleashed by the Creator. The jackal, seeing his mother's dress, tried to take it off her. And the Earth, no matter how hard she tried to hide away, was unable to avoid the incest that endowed the jackal with the power of speech. The Nummos came down from the sky to earth one day to meet their descendants and a lake was created around their vessel. This perhaps explains the Dogon's purported extra-terrestrial origins. They believed that the invention of the hoe was the beginning of agriculture, but that alone was not enough: rain was needed, and the Nummos would take care of that: "All that was impure was cast out with the water and carried away by the rains." And then the farmers began to sow.

One of the Nummos took on the aspect of a golden ram bearing a calabash that symbolized the Sun. It floated on a bed of stars and sometimes turned into a bull. It was never clearly depicted as a ram, because to portray it was to sully it or name it. This would be like forcing it into being. Ogotemmêli said that during the rainy season, before each storm, the Nummos could be seen moving across the sky.

The Dogon conceived of the world like a granary, likened blood to rain, used the same word to refer to mother and cow, and saw the clouds as the breath of life. Therefore their sacred places were oriented according to the rain, and always featured "cloud hooks", whose "curved shape retains the rain and with it the crops necessary to men". At the entrances to these shrines, among other images, horns would be painted. Whether by coincidence or not, in Griaule's drawing of the shrine, six dots are shown surrounding a larger dot, with a pair of horns over the top. He interpreted the image as that of a rainy sky.

The Dogon have always been associated with the stars, particularly since anthropologist Germaine Dieterlen wrote about certain notions of astronomy recounted to Griaule by Ogotemmêli. Their knowledge of Sirius was so advanced that numerous researchers were intrigued by how they had been able to see Sirius B without telescopes, long before western astronomers had done so. People eventually concluded that Ogotemmêli had been tainted by contact with people, such as Griaule himself, who knew about modern advances in astronomy. He was found to hold the same mistaken notions as European astronomers of the time. But even so, there is no escaping the possibility that a people with such a strong relationship to the stars and water had been depicting the constellation associated with rain by so many cultures.

Other scientists challenged Griaule when later interviews with the Dogon revealed an ignorance of the cosmogony about which he had supposedly been told. Whether they were shared beliefs or creative licence on the part of old Ogotemmêli, Griaule had made it clear that only a few initiates possessed such knowledge. We will never know what truly went on. But Griaule shows us in another of book, *Water God*, that despite our differences and prejudices there is seemingly something deep down in humans that is always the same. And perhaps that is so.

The Nummos contain almost all the elements associated with the other gods and spirits of water that we will discuss in the coming pages, and had equivalents in places we have already discussed. The Apkallu in Mesopotamia, creations of Enki, were very similar. One of them, Oannes, was depicted emptying jugs of water like the Egyptian god Hapi and the Mixtec god Dzahui. Other native peoples in Africa and the Americas still credit twins with the ability to bring rain when happy and drought or storms when angry, and believe that amphibious beings are responsible for the rain. "Be still, breath of twins," say the Tsimshian of British Columbia to ward off a storm.

The Dogon association of water with blood is also present in the vocabulary of many Romance languages. No wonder blood was part of ancient rain-invoking sacrifices, which could be performed only by certain chosen, powerful individuals. In Greek mythology, the blood of the gods had a special component that bestowed their immortality upon them: ichor. Paradoxically, any mortal who came into contact with it was bound to die. In later times, ichor fused with "petros" or stone and gave us "petrichor", the word for the smell that comes with the first rain on dry earth. Petrichor was coined in the mid-twentieth century by some Australian geologists after studying the way it arose after a long period of dry weather. Around that same time in Uttar Pradesh there were also attempts to capture petrichor—known there as *matti ka attar*—and make a perfume out of it.

Petrichor may be among the most universally beloved smells, and science has an explanation. Our ancestors, who went through catastrophic droughts, associated it with life itself, and with survival. For them it was the sign that life went on in spite of everything—just as it still is for us. Perhaps this explains the saying from Moorish Andalusia that only the tinkling of coins or the voice of one's beloved are as soothing as the sound of water. For the same reason, it soothes and lifts the spirits of the characters in Jorge Amado's *Seara vermelha* (Red Field), which begins with the final lashings of a rainstorm: "A powerful smell of earth, lifted on the air, drifting through the houses, invaded everything. Drops of water shone on the green leaves of the trees and the cassavas. A calm silence spread throughout the hacienda, over the trees, the animals, the men." The Brazilian writer's subject was the same as that of the Australian geologists: "The fear of drought, a fear that was renewed on a yearly basis, had now disappeared. […] Artur breathed in the smell rising off the earth, and a smile spread on his face once more." There is a reason why chimpanzees have their own dance when the first rains come. The smell of newly drenched earth attracts certain insects and worms, and acts as a guide for camels looking for water in the desert. Some plants have learnt that dispersing this smell sends a message to clueless insects: *water here*. They use it to lure potential pollinators.

Dogon cosmology remains a mystery, but it is neither so distant nor so strange as to warrant thoughts of extra-terrestrials. Islam, that desert religion, also sees the origin of humankind and of all living things in water. You only need look at the sky, and go back in time, for proof that Ogotemmêli's story concerns the thirst of those he acknowledged in his opening greeting, and those he did not. It is the thirst shared by the San and the people of Jaén in Spain alike when they parade a bull around to invoke rain on St Mark's Day, as if emulating the taurine cult of the god Mithras. It is present, too, in bovid worship that proliferated throughout the Mediterranean and in the Australian aboriginal myths, because it surely comes from long ago, and perhaps from the same place. It is at once the thirst of Egypt's original settlers, of the Amorites and of the Natufians; of those Ice-Age peoples who took refuge in French and Cantabrian caves with views of the water and began painting on the walls; of mitochondrial Eve and of Lucy.

The first protolanguages, created by peoples who may never have had contact, represent rain with three vertical, parallel lines. Everywhere in the myths of Africa, Europe and part of Asia, we find rain gods who are bulls, or horned men with hammers riding on the backs of bulls. In places similarly far apart, a thirsty serpent climbs up to heaven in search of rain. And in later myths, someone bangs on a rock and a spring gushes forth. With variations, this story has been shared by Australian aborigines, ancient Romans and thirsty Madrileños. It is the story of the little girl who goes out, pot in hand, looking for water, and ends up ascending to heaven, which reappears in different places among different peoples observing Ursa Major. Or a man hunting an animal, only for the animal to ascend and save itself by turning into a constellation. How is it possible for such similar stories to repeat in such geographically distinct places? Can a contagion be inferred, or the possibility that the human psyche works the same in all different places? Or is it that these myths have common denominators because they ultimately derive from the same place?

An anthropology professor, Julien d'Huy, looked at these geographically distant and yet similar stories using the same methods used to study DNA. He stripped away the elements that differentiated them and was left with the essence of the myths about the cosmic hunt featuring a hero character who ascends to the heavens in pursuit of a quarry-turned-constellation. Phylogenetics revealed to him that this cosmic hunt narrative had undertaken practically the same journey as the Middle Eastern farmers. We have been telling the same stories for tens of thousands of years, says D'Huy, who found a common "stem" in northern Eurasia some 15,000 years ago. He also found stories about rain serpents that had travelled from Africa some 40,000 years ago. The horned, plumed or winged serpent, a rainmaker associated with rainbows, is present in myths and beliefs from the Andes to the Sahara, and from Australia to Nigeria. In the same way, ancient civilizations believed in gods of water and rain that first took the form of a bull and then, at the time when their kings attained a divine status, became horned men astride this animal. Perhaps we have to wind the clock back even further to find their origin, because it may be that some of the caves in which our ancestors took refuge contain the first clues to a story that begins with the attempt to communicate with a celestial bull to ask for rain and ends with the punishment of the rainmaker who fails to gain heaven's favour.

As *Homo sapiens* spread across the world, they also spread the stories that bespoke the first astronomical knowledge, along with religious beliefs and shamanism. According to Mircea Eliade, the last of these was a common component in all hunter-gatherer cultures. Perhaps the answer to all this lies not only in Africa, but in the thirst that we have borne ever since we left Africa. A century and a half ago, a European cave began to shed light on it.

In the summer of 1879, María, an eight-year-old girl from Cantabria, ventured lamp in hand into the depths of a cave, while her father, a fossil enthusiast, remained at the entrance looking for flint tools. For a long time, Marcelino Sanz de Sautuola had ignored this cave, pointed out to him by a neighbour but to his eyes no different from all the others in the area. But now a visit to the Exposition Universelle in Paris had prompted him to start looking for prehistoric objects like the ones he had seen there. The girl looked up and shouted out:

"Papa, look! Cows!"

Above her head, floating on the cave ceiling, were bison, deer and horses in ochre tones. And handprints that looked human, one of which had been made by a child her own age. But nobody had been inside that place for 3,000 years. The father's initial reaction was uncontrollable laughter, followed by astonished silence. Soon afterwards he made some brief notes on the discovery. He did so in all self-doubt and humility; he knew himself an amateur. His only aim was to inform Émil Cartailhac, an eminent figure in the field of archaeology, that he might have found something unique, before publishing. He was rather roundabout, tentative even, in stating what he in fact believed to be the case: that these paintings were prehistoric. But in those days it was unthinkable that people in the Upper Palaeolithic could have been artistically motivated. Darwin had after all only just started to change humanity's view of itself, and there was little openness to anything like out-of-the-box thinking. Plus Darwin himself had recently described prehistoric man as a drooling savage with zero artistic capacity. Cartailhac was of the same view, and his only reply to Sanz de Sautuola's letter was to acknowledge receipt. It was only when Cro-Magnon man came to be compared with *Homo neanderthalensis*, seemingly more primitive still, that the scientists of the day granted some kind of recognition to our ancestors. But to speak of "art" was still a step too far. Yet that was precisely what Sanz de Sautuola did, if with quite a bit of preamble. It took him many pages—in which he even conceded the unlikelihood of what he was about to say—to arrive at his proposition: "It has been proven that man, when his dwellings were no more than caves, knew how to reproduce with considerable accuracy on elephant tusks not only his own figure, but also that of the animals he saw."

He paid a high price for those words. People had already had enough of hearing that they were monkeys, the cousins of gorillas and chimps. For those who accepted evolution, and precisely for that reason, such sophistication was simply inconceivable in such ancient peoples. There was no way babbling, loincloth-wearing cave-dwellers who ate with their hands could be forerunners of Matisse. Although people's curiosity was initially piqued in Spain, Sanz de Sautuola was discredited and ostracized by the scientific community, the Church and the press alike. The first of these, particularly in France, thought the whole thing a hoax. Some accused him directly of fraud, while others, not doubting his good intentions, none the less thought him the victim of a clerical conspiracy orchestrated by the Spanish Jesuits to discredit evolutionary ideas. To the scientific community he was just a rich lawyer with too much time on his hands—a dilettante bold enough to wade into the already heated debate between creationists and evolutionists. But Spanish

Catholics were also unwilling to accept his ideas. The only public support he received in his own country came from a journalist and also from Juan Vilanova y Piera, a palaeontologist and geologist who, despite his creationist ideas, was convinced of the authenticity of the Altamira paintings. In France only one archaeologist, Édouard Piette, dared to defend Sanz de Sautuola publicly; he was widely ridiculed for it, and he too eventually distanced himself.

But then, at the turn of the century, paintings similar to the Altamira ones were found in caves and rocky shelters in southern France. Following the discovery of the La Mouthe cave, in 1902 the prehistorian and abbot Henri Breuil gave his seal of approval to the paintings they contained. Cartailhac, who had once walked out of a room in disgust while Sanz de Sautuola and Vilanova y Piera were discussing Altamira and had published articles in his own magazine ridiculing the Cantabrian and refused an invitation to visit Altamira, ultimately had no choice but to retract his statements. In reality, he had done what is expected of all scientists: he expressed doubt. But he also allowed himself to get carried away by a scare story he had heard shortly before Sanz de Sautuola's letter, when he had been warned that Spanish clerics were preparing a trap for evolutionism. And the combination of fear and scepticism led him to overreact, while his elitism meant he missed the opportunity to open a door that an amateur had left ajar. To make amends, he penned an article in which he pleaded guilty to the injustice suffered by a man who had in fact been in the right. Finally, he travelled to Altamira together with Breuil and there found what he believed to be "the most beautiful, the most strange, the most interesting of all the caves with paintings". They say that he apologized to María, who was no longer a child. But he arrived too late: her father was dead by then.

So it was that highly similar paintings came to be found across southern France and northern Spain. The people who painted them had been there between 30,000 and 15,000 years ago, sheltering from a frozen, arid world. During that period, *Homo sapiens* may have crossed paths with the last remaining *Homo neanderthalensis*, who had already become extinct in all other parts of the world. But soon they were definitively alone, and the Iberian peninsula and southern France became the final refuges for those fleeing the cold and aridity of the last glaciation. There the new arrivals set about drawing, although a kindred artform was being produced almost simultaneously in Australia and Indonesia, and indeed had been produced far earlier in Africa.

Who were these cave artists? Did they go into the caves because they were cold, and did they start to draw out of boredom? Were they compelled to represent animals in motion, or to show their gratitude for enabling them to go on living, or was it their way of invoking the game they hunted? All this and more science now began to ponder. Paradoxically, it would be clergymen like Breuil who would go on to champion the study of paintings that at the time were so at odds with both creationist ideas and prevailing scientific opinion, both of which thought prehistoric *Homo sapiens* capable of little more than standing upright when awake.

In Montignac, a village in the Dordogne in southern France, much the same happened as in Terrinches. Somebody stopped for a moment to listen to the old folk, who spoke of a treasure hidden underground, and wondered what truth might be contained in stories like these that had been passed down through the generations. They concerned a cave in the vicinity of Lascaux Manor, from which sounds were said to be heard. After 1920, when the roots of a tree felled in a storm left a hole so gaping that a donkey fell down inside it, legends and speculation became increasingly rife. Shepherds covered the hole to keep their animals from falling inside. Twenty years later, Marcel Ravinat, a teenaged mechanic's apprentice, went there with his dog, who was called Robot. The dog began to dig in the place where a small hole had been left, and his owner followed suit. But it was about to get dark and, finding it far deeper than expected, Marcel decided to go home.

He came back a few days later with three friends. Armed with a penknife and a homemade lantern, he hauled brambles aside and commenced digging, then crawling, only to fall into a pit, dropping his lantern and finding himself in pitch darkness. His friends went down after him. There was space for the four of them and much more besides—the lost dead donkey included. When Marcel found the lantern again, they looked around to find figures clearly depicted on the walls: horses, deer, aurochs. They had just discovered, with Robot's inestimable assistance, the grain of truth in those legends, because the cave in which an abbot was rumoured to have hidden during the French Revolution indeed existed, and so did the treasure.

When they finally decided to share their discovery, the boys took it to the town's old teacher, who in turn wrote to a teacher, abbot and archaeologist who was in the area at the time. This was Henri Breuil, who had left Paris five months earlier, fleeing the war. He was by now an authority on Palaeolithic cave art; not a single discovery happened in Spain or France without his making an appearance. The same Breuil who had given his approval to Altamira, and whose sketches of the bison there inspired some of Picasso's work, had been given the nickname of the Pope of Prehistory. When the old Montignac teacher told him about Lascaux, he was in the right place, barely twenty-five kilometres away, at the right time.

Approximately 17,000 years ago, someone went into the deepest part of the cave. Using brushes made from plants or bits of bundled-up wool, and with iron manganese minerals and carbon clay, this person proceeded to depict a huge auroch—a wild bull—facing the left-hand side of the chamber. Dots were painted around the creature's eye. Except for the tilt of the horns, the picture is almost identical to the one in the Dogon temple that Griaule considered a representation of a rainy sky. Lascaux and Altamira can perhaps be considered proof of our deep relationship with the skies ever since the Palaeolithic Era. Although nobody saw it at first, it could be that they also speak to our place of origin.

Some researchers put it down to boredom. We now know that long before, even *Homo neanderthalensis* had an artistic sense. But at the time these paintings were discovered, scientists thought the bull could at most be an artistic expression without symbolic or ritual value. It had already taken a lot for them to accept that humanity's first artists were those they had dubbed slobbering savages. Henri Breuil, however, found cave art to be endowed with a religious meaning, and he was possibly the first person to intuit the spiritual value of those animals repeated on either side of the Pyrenees. Save for a handful of archaeologists, such as Alexander Marshack—who had already begun to speak of Palaeolithic astronomy—virtually no one considered the possibility that such ancient peoples could have been concerned with the stars or sought to reproduce them, or had any kind of ambition beyond painting for painting's sake. "Archaeoastronomy"—the investigation of the astronomical knowledge of prehistoric cultures—had not yet been invented.

HEAVENLY HORNS

But when studies of the Hall of the Bulls began, it became clear that the paintings bore a striking resemblance to the night sky, and that the animals had been painted at what would have been the mating season. The idea that someone had painted the animals to mark the appearance of particular constellations that allowed them to tell the seasons was a controversial hypothesis. But from the early 1980s onwards, the German astronomer Michael Rappenglück couldn't stop asking himself a question that became the foundation of his doctoral thesis and his life search: what if the Lascaux paintings were maps of the stars? He compared the dots floating over the bull's back with the cluster that makes up the Pleiades—not as the constellation is nowadays, but as it would have been at the time. Now it can be seen from April to October, and although this hasn't changed that drastically, Rappenglück established what its exact cycle would have been 20,000 years ago. The Pleiades would have been at their highest in early spring and, after a few months of shining brightly, would have disappeared at sunset on 26th August, reappearing again on 11th October, after the autumn equinox. Continuing his search, Rappenglück found aurochs also suggestive of the Pleiades in other caves of the same era, and by gathering myths about the constellation from cultures around the world, he came to consider the Lascaux auroch part of a calendrical system that divided the year according to the moment the Pleiades became visible—which heralded spring, rains and hunting—and then went away again. And what about the dots around the creature's eye? "If the hypothesis is correct that the first set of dots floating above the animal represent the Pleiades," he wrote, "one can assume that the second indicates the other cluster of stars in Taurus, the Hyades, distributed around the main star of that constellation, Aldebaran." According to the German astronomer, it may then be that the worldwide belief linking the appearance of these stars with the arrival of the rainy season first arose during the Palaeolithic. He believed that it might have marked the beginning of the year as such, and that this was intimately related with the auroch mating season, given that the indigenous Blackfoot people of Montana and Alberta synchronized the Pleiadic phases with the buffalo's natural cycles there, which also provided the basis for the naming of the months in the Sioux and Cheyenne calendars. Rappenglück however found that links between the Pleiades and bovines were not universal, but might have spread far later on from Mesopotamia and throughout the Mediterranean and the Indian subcontinent.

Researchers went on looking for stars in the rocks. In Lascaux, they even detected other constellations and what could have been a warning about a forthcoming comet shower. Ideas connecting rupestrian art with the night sky went beyond just the French caves. The Spanish artist Luz Antequera put forward the possibility that the Altamira bison also contained a message about the stars and the weather, which was not exactly hidden but that nobody had previously considered. This was a daring proposition, bearing in mind the precedents, but the fact that they were painted on the ceiling and shared certain attributes with ancient Egyptian representations of the sun led her to wonder whether it was a prehistoric image of the celestial vault. This would not be strange if we bear in mind the ancient belief that stars and water come from the same place, and the fact that Babylonian priests and astronomers saw the Pleiades and the Hyades as, respectively, the fur on the back of the celestial bull and its jawbone. Could these paintings have been the first example of the gods being petitioned for rain? Might the origins of rain-magic lie in Lascaux and Altamira?

✿

There are those of us who look up at cloudy skies and see dragons and dinosaurs, even animated sequences. Our minds draw without the use of a pencil. The doors and windows of a house can form a face on which we even confer emotion. It seems that our brains process these non-existent shapes just as they do real ones. The phenomenon has a name to match its beauty: pareidolia. We now know that precisely the same thing happened with our most ancient forebears. There were nights when they looked up at the sky. What would they have thought those lights out there were? Distant encampments, is Carl Sagan's view. They would have seen that the rain, the light, the warmth and even the lightning bolts, which might have enabled the discovery of fire, came from there. That was why they started to pay more attention to the stars, which for some time they still thought of as lights rather than something existing on a separate, astral plane.

Looking at them must have been like seeing one's home in the distance and not being able to go back. This might have been the moment when, stopping in their tracks to gaze calmly at the sky, they experienced nostalgia for the first time. They would not have known that, long, long before, the recently formed Earth had been so hot that water in liquid form could not have existed on it. The first stars had to die in order for water to arrive on our planet, and from it they themselves perhaps emerged. Those people who observed a razed home with a single light left on surely took from the stars something as profound and invisible as the calm ushered in by the smell of wet earth. Perhaps astronomy, religion and the calendar, which still govern our lives, are just different forms of thirst. The luminous pinpricks in the celestial vault would not always have presented the same picture to them. By starting to make use of throwing weapons, and by observing the changing skies, they would have become aware of time, distance and their own mortality. They would have asked themselves certain essential questions: where did all of this come from? How did we get here? What is that thing which sometimes falls from the sky? The answers were contained in the earliest myths. The primitive cults, orientated around the dead and fecundity, would inevitably have been connected with the heavens and with the Earth's cycles of renewal. A system for counting time was created, marked on bones. And the calendar had been invented, possibly in the Dordogne. There, on a limestone slab very near to Lascaux, someone carved a naked woman holding up a bison's horn. The Venus of Laussel has been associated with the lunar cycles and with fertility, but in her hand she held something that had already been widely depicted in horns of plenty.

In reality, this was all a slow process. It was approximately 1.5 million years between the creation of throwing weapons and the oldest recorded lunar calendar.

Those people's imaginations joined twinkling lights together, traced sidereal lines, drew animals in the sky. The first shapes that stood out to them were the figures of a bull and a lion. These were the first constellations, although there would be no written proof of them until the Babylonian astronomers created the so-called MUL.APIN compendium, in which Taurus figured as "the stars". But *mulapin* also meant "plough", which constellation marked the Babylonian new year and the great innovation that replaced manual tillage. With the help of oxen, that invention made life far easier for the human beings who had—for reasons still under discussion—chosen the hardest, most laborious of paths. Plough and bovids would be present from that moment on in stars, myths and religious beliefs. Why the bovid? For Mircea Eliade, many innovations from the Neolithic to the Iron Age found echoes in religion. That is, the bovid's prominence was due to the invention of the plough—but how then to explain the Lascaux auroch, which massively predates the invention of agriculture? At that time, the first proto-calendar was already obsolete because human beings needed a more detailed understanding of the workings of the periodic renovation of the world, on which their food, and the early glimmerings of their faith depended. If they wanted to guarantee something as volatile as a harvest, they would have to learn to manage the rain: by invoking it or directly requesting it from the relevant party, and, if possible, predicting and retaining it too.

We can only wonder if it was from the Neolithic onwards that the presence of rain began to signify abundance and life, and its absence scarcity and death. For hunter-gatherers in the Palaeolithic, the auroch was a symbol of fertility and abundance, possibly because the earth became bounteous just at the moment its likeness appeared in the night sky (which was also at the beginning of the auroch mating season). They may have realized already that the auroch also accompanied the rains, but their descendants, the sedentary farmers of the Neolithic, would surely have viewed this bovid quite differently—as a draught animal and provider of milk, cheese and leather. It also gave them the signal that the rains were coming to fertilize the soil and ensure their harvests.

With agriculture also came the myths explaining its own origins. Stories and the act of storytelling would have been great fortifiers, for the way they enabled an understanding—and therefore the overcoming of the frequent threat of droughts and floods. Such shared tales would also have fostered cooperation within groups, which was increasingly necessary. So we can see that, over the course of thousands of years, thirst took up residence in myths and rituals. And many of them had a bovid of some description in a leading role.

When they looked up at the night sky, inhabitants of the Fertile Crescent, many of them climate refugees, saw the pictures there beginning to move, and then those motions repeating. Plotlines were soon fitted onto that cosmic pareidolia to make sense of them. The heavens housed a story that served in equal parts as agricultural calendar and a basis for religious beliefs. Contemplating the heavens, people witnessed what was perhaps the first ever motion picture. The plot—spoiler alert!—went something like this:

The story began 5–6,000 years ago, at the time of year when Taurus and Leo swapped places in the night sky. Taurus, which had dominated the winter nights, would cease to be visible at sunset on 10th February, when Leo assumed its highest position. A confrontation between bull and lion would have been clear to see; when the feline won and took over, the bovid died and disappeared. Humans would then seize the moment to plough and sow, in hopes of a year of plenty. But forty days later, the bull would mysteriously revive, coming back to take vengeance on the lion. Then the rains arrived. Down below, on earth, the first shoots would begin to appear.

The bull and lion constellations indicated the most important moments in the agricultural cycle, and it was a plot very similar to the ones that pitted Seth and Osiris against each other, not to mention Baal and Mot, Tur and Iraj and Huang Di and Chiyou. That Jesus's time in the desert, now marked by Lent, lasted forty days, and roughly overlaps the dates when the celestial bull is absent, is also surely no coincidence. Other explanations also allude to thirst, and to the same number: the forty days of the Great Flood and the forty years of the Israelites' Exodus. And Jesus wasn't the only one who spent forty days fasting in the desert; Old Testament prophets like Moses and Elijah did the same. Curiously, forty days was also the period during which the bull chosen by ancient Egyptians to represent Apis was kept hidden from view, except to a number of women who would have disrobed before it. Once that time had passed, and coinciding with the full moon, the beast would have been transported to Memphis for coronation.

The film in the sky, in which the bull is water and life and the lion is thirst and death, was immortalized in the world's first known epic, that of Gilgamesh. Gilgamesh, a tyrant king, is visited by Enkidu, who has been sent to kill him. Against all odds, the pair become great friends and set off together in search of adventure. The Mesopotamian goddess Ishtar (Inanna to the Sumerians) becomes obsessed with Gilgamesh and asks to marry him, but knowing how she has treated her previous partners, he spurns her. Among a litany of compliments, he describes her as "a brazier that goes out when it gets cold", "frost that does not become ice", "a door with slats that keeps out neither breeze nor drought" and "a shoe that crushes its wearer's foot". Reminding her of her treatment of past lovers, he asks: "Which of your lovers lasted forever?" Ishtar is so enraged that she ascends to the sky and asks her father, Anu, to let her have the bull of heaven so she can destroy Gilgamesh, Enkidu and everything else besides. Gugulana, a deity from Sumerian mythology, represented the constellation of Taurus. In Sumerian, her name means "great bull of the sky": *gu* ("bull"), *gal* ("great"), *an* ("sky"), *a* ("of"). Anu refuses at first but, in the face of his daughter's fury and hellish screams, finally gives in. Ishtar wants Gilgamesh to die of thirst, along with anyone else who gets in her way. And Anu says to her:

> *If what you want is the bull of heaven,*
> *the widows of Uruk must gather straw for seven years,*
> *and for seven years [the field workers of Uruk] must make hay.*

Ishtar is undeterred. She knew this was how it would go, and has warned the widows and workers of the coming drought:

> *[Down came] Ishtar, leading it onward:*
> *when it reached the land of Uruk,*
> *it dried up the woods, the reed-beds and marshes,*
> *down it went to the river, lowering the level by seven full cubits.*

A devastating seven-year drought ensues. Every time the bull of heaven tramples its way through Uruk, a grave opens up and hundreds of men fall inside—Enkidu among them, although he manages to clamber out at the last moment. Gilgamesh and Enkidu battle the thirst-bringing bull, and the former finally kills it, plunging the knife in while his friend holds it by the tail. Enkidu tears off one of its legs and flings it at Ishtar. On departing from Uruk, Gilgamesh shouts at the onlooking women that he is the best-looking man of all, and goes off to celebrate in his palace.

When agriculture produced its first great innovation, the plough, the ox proved to be one of the first friends to humankind, who had not long before domesticated wolves to turn them into dogs, as well as the grasses that would give them cereals and the mouflons that would later become sheep. The choice of oxen to plough the land, though a very different treatment from that of castrated bulls, none the less surely reinforced the overall sacredness of bovine creatures—a sacredness acquired in the days when aurochs still roamed the earth, when they were associated with fertility and abundance, just as the cow was also associated with the mother goddess in some cultures. The film of the sky also gave bovines a new connotation: as opposed to the lion, which represented death, the bull was a symbol of life.

Perhaps because of its importance in the Mediterranean, it became a divinity there too. Altars in the shape of cows proliferated, as did representations of bulls and horns in the caves (typically these were side-on, facing left; the same orientation as the Taurus constellation). This was also the case in Anatolia, the north of the Iberian peninsula, southern France and some parts of Africa.

When the Phoenicians invented the alphabet they would go on to bequeath to us, they created a symbol for the ox already present in the hieroglyphics of Egypt—whose culture venerated the bull-god Apis and considered cows sacred. So important was it in their lives that they placed it foremost in their alphabet, as was also the case in the Proto-Sinaic alphabet. If we turn our letter "A" upside down, we see horns, although originally the letter 'āleph (✦) looked more like a bull in profile. *Alp* was their word for ox, and it is the reason our alphabet begins with this letter, although the shape differs slightly.

In Anatolia the domestication of the bull marked a radical shift. Bull worship very likely spread from there, leaving a trail all the way to Tartessos (modern-day Huelva, Cádiz, Seville and Badajoz), and to Africa as well, where traces remain to this day. Thus humans began placing bovids at the centre of both their physical and spiritual worlds; at the same time as domesticating and putting them to work, they prayed to them for abundance and honoured them with a special place.

Over the course of several days in the 5th century BC, the inhabitants of Casas del Turuñuelo, in Spain's Guadiana basin (Guareña, in the province of Badajoz), held a great feast, lit a fire, piled up the meeting place with mud and sealed it all with adobe, turning it into a burial mound. Shortly afterwards, and quite suddenly, they departed northwards. Before the journey, a great public sacrifice took place: dozens of animals were slaughtered, horses especially, but also a number of bulls, pigs and one dog. Such hecatombs were carried out in ancient Greece as a way to appease the gods, originally with 100 oxen, although over time other animals were added in and the number ceased having to be exactly 100. Why were they trying to appease the gods? Surely prolonged droughts or floods must have been among the reasons, although researchers at Casas del Turuñuelo have tended to see it more as the latter. But the sacrifice seems not to have worked, and the people there moved away instead.

Thousands of years later, alongside the remnants of the sacrifice and the banquet (around 200 plates, twenty wine glasses, bowls, platters and barley seeds that nobody cleared up), archaeologists found altars featuring bull-shaped carvings.

Tartessos was once dismissed as pure mythology. Philologists were among the first to try to establish the truth of it, although they thought it was a city they were looking for and not a culture. Then, in the mid-twentieth century, the Carambolo Treasure was discovered near Seville and Cancho Roano. In 2015, a chamber was found in Casas del Turuñuelo containing a bull-shaped clay altar associated with the god Baal, who was Canaanite in origin but shared by various peoples, the Phoenicians included. There was also a staircase and, two years later, a courtyard with the remains of the hecatomb was discovered at the bottom.

Although the identity of the Tartessians is not clear, the consensus tends to be that they were mixed descendants of Phoenicians and local peoples. We know of the existence of the culture in the Guadalquivir valley from the 9th century BC onwards—coinciding with the arrival of the Phoenicians in the region. In the 6th century BC, however, a serious crisis, seemingly triggered by a drought, drove them north, particularly towards the Guadiana, in modern-day Extremadura. This final act would not last more than 100 years. By the end of the 5th century BC, Tartessos no longer existed.

Around the 7th century BC, depictions of a conflict between a bull and a lion, just like the one in the sky, began to appear in places from Phoenicia (modern-day Lebanon) to Tartessos. Various archaeological finds in southern Spain, Portugal and Turkey have shed light on our ancient relationship with bulls at a time when they were sacrificed on altars—which were themselves bull-shaped—as a way of invoking the rain. In sites as far apart as Göbekli Tepe, Çatal Hoyuk, Carambolo, Coria del Río and Cancho Roano, these bull-shaped altars, seemingly Phoenician in origin, became widespread, as did horns hung on walls. And at Cerro de la Encantada, a Motillas site in La Mancha, an altar was found; it has been called "consecration horns", an image extremely common in the Minoan culture of ancient Crete and based on a sacrificed bull. As for the Çatal Hoyuk horns, possible meanings have been found linking them with both astronomy and petitions for rain. In Canaan, the story that everyone saw in the sky was already being told: Baal, as god of storms, and Mot, as god of droughts, faced off in an endless cycle; Baal was the bull and Mot the lion. Sometimes the bull's victory was represented as a way of compelling the heavens to intervene and send rain. Five or six thousand years ago, the nights when the bull of heaven shone brightest would have coincided approximately with the days around what is now 15th May. In terms of humanity's search for water, a not insignificant date.

Ishkur, Adad, Hadad, Baal (or Bel for the Akkadians) and Teshub were all gods of the same thing. Everything points to the fact that the cult of the bull, or rather the divinity it represented (rain, thunder, fertility), has its origins in the East. Storms and rain were embodied in Sumerian, Akkadian, Syrian-Palestinian, Canaanite, Phoenician and Hittite divinities. Due to their association with fertility, these gods were usually linked to the corresponding mother- or land-goddess, such as Inanna/Ishtar, Astarte and Khepat. This idea is strongly present in the ancient religions of the Fertile Crescent and in much of the Mediterranean. For Scottish anthropologist James Frazer, these beliefs are a way of representing reproduction itself, with water being masculine and earth feminine. Marija Gimbutas, meanwhile, saw the old European cult of the bull as being linked to the relationship between horns and the moon, which would explain why so much emphasis was placed on the horns; these would then be a symbol of the moon, and the primordial sacrifice of the bull seen to give rise to new life. This bovid was used to ask for rain or fertility. The bull also often represented the god of storms—because its bellowing reminded them of the sound of thunder? Elsewhere, we more often find beliefs that associate snakes with water, rain and clouds. However, as the story of Ogotemmêli shows, at least in some parts of Africa the two approaches coexisted and merged. Evidence shows that this has been and continues to be the case.

In the past, in other parts of the world, it would not have been uncommon for a bovid to be sacrificed as a way to put an end to a prolonged drought. Frazer tells us that for the Egyptians, oxen "above all others had helped the discoverers of corn in sowing the seed and procuring the universal benefits of agriculture". In Attica, as Frazer also recorded, oxen were sacrificed and their flesh ingested, while the hide was stuffed with straw and the resulting effigy set to ploughing the dry fields as a way of conjuring rain. Then a trial would be held to determine who had killed the sacred creature. The blame was passed from one to another, until finally the instrument that had dealt the death blow—an axe or knife—would be accused, and thrown in the sea. On the Gulf of Guinea, oxen were made to cry by having flour and wine thrown in their eyes, while everyone chanted in unison: "The ox will cry! Yes, it will cry!" And if tears did come to the animal's eyes, it was taken as a portent of rain.

For the /Xam nation, *!khwa* is the rain, a living being that sometimes turns into a bull. It lives inside deep puddles, but when it leaves the puddles of its own volition, they dry up; the rain bull is gone. The job of the shamans is then to go out and catch *!khwa*, which, in its bull form, is paraded across the dry earth before being sacrificed. After trampling the bull's flesh, they throw chunks of it in the places where they want the rain to fall. But it is only possible to catch *!khwa* at the right moment. One of the /Xam myths tells of a hunter who catches *!khwa* by accident, while it is in the form of an eland. The people try to eat it, but its flesh is consumed by fire, and all the members of the community are turned into frogs.

Since the sixteenth century, in Spanish and Portuguese villages across the south of the Iberian peninsula, a man has always gone out into the countryside to catch a bull on the eve of St Mark's. He has to use a specific set of words, a kind of conjuration, to make the bull go with him back to the people. The villagers, believing the bull magically tamed, take it in procession along with the saint, scattering it with flowers and then bringing it inside the church to attend Mass like one more parishioner, afterwards to be sprinkled with holy water by the priest. In some places, this ritual coincides with rain supplications. Then something far from unexpected takes place: the bull regains its vigour and finally goes back out into the fields. In fact, it isn't that much of a miracle, as Vicente Moreno related after witnessing the ritual in 1927: "The young bull tries to escape, but all the men around beat it with clubs and sticks, before, with the priest leading the way, it penetrates the church and, passing along an avenue formed of the faithful, goes up to the main altar, before returning the same way back onto the street." The ritual of the noosed bull is in essence very similar to that of the /Xam nation, and may have Neolithic origins.

||Kabo was a San rain-shaman who summoned the rain in his dreams. One day he felt a pain in his arms and chest and asked the rain to come. Then he fell asleep and spoke with it.

In the middle of the Kalahari, that desert in which we may all have originated and in which members of the San still live, there is a place called Tsodilo to which our distant relatives still go. These are sacred rocks in which, the San say, the spirits of the gods reside. In the early twenty-first century, some 4,500 cave paintings older than those of Lascaux and Altamira were found inside these caves. And they have elements in common with the French and Spanish ones, as if they were all by artists of the same school. The stars of the show here are animals, sometimes horned; people, if they appear at all, are little more than background figures. It was a time when the gods of fertility and rain were horned animals, not men with beards and horns. The San call that place "the mountains of the gods and the whispering rocks". The rest of the world, on the other hand, knows the first museum of humanity as the "Louvre of the desert", as if it would be wrong to call the Louvre the "Tsodilo of Paris". From there, after all, people emerged who went on to leave their mark in caves on every single continent over tens of thousands of years.

HEAVENLY HORNS

When the scientific community recognized the prehistoric paintings of southern France and northern Spain as valid, it was practically as though cave art existed there and nowhere else. Although other cave paintings appeared in places like Australia and Indonesia, until recently it was believed that *Homo sapiens* must have been the first to pick up a brush in Europe. But at the same time that people were painting on the French-Spanish borders, so too were those who had remained in Africa, as demonstrated by the San paintings dating back 25,000 years.

The artistic side of *Homo neanderthalensis* has recently been discovered too, given the evidence dating back to a time when there is no proof of the presence of *Homo sapiens*. In South Africa, paintings and engravings have been found that are up to 73,000 years old. If we wind the clock right back, we end up in the place where everything began. If we look for what it was that united *Homo sapiens* and *Homo neanderthalensis* long before they met, we return to Africa, the home of their common ancestor. If both painted walls, we cannot be certain that *Homo ergaster* or *Homo antecessor* lacked the same ability before leaving that continent. Tens of thousands of years later, their descendants still hold sacred a set of caves teeming with painted animals that are, to them, the very spirits of their gods. Our African ancestors had surely already observed both the sky and animals while searching for answers to their first transcendental questions. If we could avoid being blinded by ethnocentrism and anthropocentrism, we might be quicker to find the answers to certain questions, and perhaps a way of moving beyond recurring problems already resolved by our ancestors long ago.

7

God came to earth

> In another time, there was no rain. The people lived on a dry earth. Only dry soil, only hard rock, hard soil. The plants, the animals and the men who needed the rain went and offered prayers to a swamp (*no'yo*), and vapours started to appear from a mountain peak, named *nu ñu'un no'yo* (god of the swamp) or St Mark… From those vapours the clouds formed, eventually "maturing" to fall in the form of rain.
>
> MIXTEC ELDER IN *WE ARE THE PEOPLE WHO EAT TORTILLAS* BY JOHN MONAGHAN

During my childhood, every time 25th April came around, I tied the horns on the devil. It isn't as epic as it might sound; we just took a few stalks of barley or corn, still green, and braided it. All the families in the village went out into the fields together, and there was the culminating ritual of eating the *hornazo* or Easter pie, to which my grandmother added chorizo—of course—even though it was a sweet. Not really understanding what was going on, I imagined a lord of the underworld with green hair trapped under the earth, and that it was my fault he was unable to give forth his shoots. But the grown-ups said that was a good thing and I, as a little girl, followed their lead without too many questions—really I was interested in getting to eat the *hornazo*. The point of the ritual, as I now know, was to stop the devil from hexing the harvest, which is like invoking timely rain. Asking it to come in time, asking it to carry death away. Mark, along with Isidore, is the saint most petitioned for rain, their festivals coinciding with the beginning of the agricultural year; the former is possibly the incarnation of the heavenly lion, the latter the bull, as their iconographies suggest. In popular lore, St Mark came to be known as king of the ponds long before crossing one huge one—the Atlantic.

Nowadays, in certain parts of Mexico, St Mark is known as Savi, which is both a word for "rain" and the Mixtec name for their principal God. He is remarkable for his similarities with Hapi and Enki, but also because few people embody thirst-inspired beliefs in arid places like the Mixtec do.

In the beginning was chaos. The earth was an emptiness, devoid even of time. In the darkness, a Jaguar-Snake and a Puma-Snake flew about. Then one day Jaguar-Snake descended to Earth and became a man, and Puma-Snake fell, taking the form of a woman. This primordial couple engendered the *ñuhu*, who were the planet's first inhabitants; they were gods of rain, air, Earth and moon, as well as of predictions, mountains and animals. They settled on a riverbank and began working to create the first humans. When the sun appeared in the firmament, the *ñuhu* fled inside some caves, where they were turned to stone. One day, the god Dzahui left his cave, journeyed to the heavens and returned with a jug in his hand; he then began to anoint the future king with water.

Hence the Mixtec belief that Dzahui lives in a cave, is the rainmaker and becomes manifest through droplet-shaped stones. Carvings show how his face is imagined to be: moustachioed, fanged, and with a sort of goggle mask or blinkers on. When they find a stone apparently out of place, they ask Dzahui's permission to move it, and if he grants it, they take it to the stone yard, situated atop a "rain ridge" specifically dedicated to invoking the rain. The stone then becomes an object of worship. The wind, the Mixtec believe, visits the caves looking for vapours to transform into clouds. It loads up the clouds and carries them away while they continue to mature and are eventually set free in the form of rain.

Those who venerated Dzahui called themselves *ñuu savi* ("the people of the rain") and lived in the historical Mixteca, in southern Mexico (in parts of what are now the states of Oaxaca, Guerrero and Puebla). In their language, the territory is called *Ñu Savi*, the "land of rain", but in the fourteenth century the Aztecs renamed it Mexicapán ("land of clouds"), and the Spanish later called it Mixteca and dedicated it to St Mark.

Dzahui is worshipped in concert with Snow Wind, also known as Rain Snake: the hurricane. Some Mixtec believe that the plumed serpent lives in the clouds, while others hold that it emerges in the month of May and then lifts the clouds up higher, moving them around and bringing the rain. The plumed serpent is a harbinger of rain, as are dust devils. In Mixtec culture, clouds and smoke are identical, so in times of drought shamans invoke rain by going up to a hilltop to smoke tobacco.

Every 31st December, at midnight, the Mixtec wait for the first cloud to appear, watching for its direction of travel. If moving south to north, it portends fair weather; if north to south, there will be a drought. Worst of all is if no clouds come. And this isn't superstition: northbound clouds do in fact bring rain here, while those heading south do not. Those who live in the south of Mixteca have heavy rainfall to contend with, but the north is semi-arid. For this reason, in the north, where they live on subsistence farming, the cult of Dzahui predominated from the 5th century BC onwards.

The Mixtec sense of time is determined by the alternation of rain and drought—that is, of life and death. They believe that birds bring the rain, just as the Chumash people of central and southern California believe ravens specifically bring rain, and the ancient Egyptians believed it to be the work of herons. By St Mark's Day, the rainy season is underway, lasting until around All Saints' Day, when the dry season returns, along with the dead. For them, the deceased are seeds, some of which turn into rain deities, since they call plants "sons of the rain".

The connection between thirst and death in myths and beliefs can be seen in places so far apart that it can perhaps be considered universal. "Thirst" comes from the Proto-Germanic verbal stem *thurs-*, which in turn is from the Proto-Indo-European root *ters-* "to dry". Everywhere rain gods tend to be found in constant battle with the gods of death, who are usually also the gods of thirst. Dying, after all, is very much like drying out. The water in our bodies decreases as the years go by, and we also become less thirsty towards the end of life. We are little more than a drop that evaporates before becoming another drop. Rooted in our unconscious, the fear of dying of thirst, or of the dead not having enough to drink, has always been with us. In Mesopotamia it was important to look after the dead so that they did not leave the earth and become a danger to the living. Their bodies were taken along even when people were forced to migrate. Anyone who died without a grave or some kind of care taken over them was especially feared.

"I thirst" are the two penultimate words uttered by history's most famous dying person. Although it wasn't water that Jesus wanted, this is among the most common requests in the final hours, although it's often those around the deathbed who make it. It seems that we die with the same necessities with which we are born, but the body of a dying person no longer really needs water. As something that more properly obsesses the living, the placing of some water vessel alongside tombs is a time-worn tradition, as we have seen. In some areas of Mexico where Dzahui and his counterpart Tláloc were worshipped, a glass of water is still placed next to the deceased during the wake in case the person should suddenly wake up thirsty.

Although Mexico had various rain gods, and that went for the rest of the world as well, few of them were as elevated as Dzahui. Baal possibly was; worshipped alongside other gods by various peoples of Asia Minor, he ended up proclaiming himself king of the gods in Mesopotamia. The resemblances between Dzahui's and Baal's innumerable relations are clear.

It might be that it all began with the bull of heaven in Mesopotamia and with Apis in Egypt. The Egyptians held oxen as sacred for the assistance they had provided in turning grasses into cultivable cereals. Apis was particularly prominent—initially associated with Osiris but finally becoming a god in his own right, and going on to be adopted by the Greeks and Romans under different names. But at a certain point, bovid began giving way to human, and the rain gods, like Apis himself, were for a time half man, half bull. The Baal of the Canaanites had the body of a man and the head of a bull, although in later depictions by other peoples he appears as a horned man. The Phoenicians depicted Baal (also known as Aleyin or Aliyan) with slender horns. He was responsible for bringing rain to the fields. He had to vie with Mot on a cyclical basis; the latter made the soil so hot that sometimes Baal had to be petitioned to bring the wetness back. But at the end of the driest season, Baal always ended up defeating Mot. He had the help of Anat, who was both his sister and wife, and the goddess of war and love to boot; she would go out and spread dew on the dry soil. Mot would then die, but only for a few months. And although he always came back to life, he was also the God of the dead and of sterility. Baal had an equivalent among the Asturians and Cantabrians, Candamius (synthesized with Jupiter, he could also have been the Celtic god Taranis, "the thunderer"), who spent most of his time in the mountains, surely close to the *nuberus* of local mythology. The *nuberus* were a kind of imp, always depicted wearing a hat, and they were in control of the clouds and the rain. They had an equivalent in Las Hurdes (Extremadura) called Entiznáu, although he was more of a giant who reached up to the clouds and made it rain by stirring them— with his hat. Entiznáu also had the ability to summon lightning with a stone and flint like the one my great-grandfather used, and which I claimed as a keepsake long before I knew anything about this giant.

Dionysus, the Greek god of agriculture and grain, was born with horns and, from the time he was a baby, would shoot lightning bolts from his father Zeus's throne. Zeus, the god responsible for rain, also had a peculiar relationship with bovids; see the episode when he disguised himself as a bull in order to seduce and kidnap Europa. Just as children tend to resemble their parents more and more as they grow older, Dionysus was not only depicted as a horned man, but sometimes borrowed his father's disguise and took on the aspect of a bull. And since he was said to have died while occupying that form, Dionysiac rituals commonly included the sacrifice of a bull in his honour. His followers credited bovids with at least having made life in the fields easier for humans during ploughing season. But, in fairly short order, the animal that had once been the god of rain became little more than a conveyance for the figure now in charge of rainfall; a man with godlike pretensions. The story of bull-adoration, just like that of other once-prized animals, ended in the usual way: after the idealization of wildness, domestication brought monotony, and a consequent devaluation of "beasts of burden". The primordial animals of the Palaeolithic began to be substituted by the men of the Neolithic. Wishing simultaneously to be the lion of heaven as well, men decided that they would take over the sacrificing of bulls to invoke the rain.

And where do women figure in all of this? Although less well known, in some places they may have come first. Gimbutas believed that in the Europe of antiquity there was a prevailing belief in a mother goddess who was eventually ousted by the god of thunder. She had the appearance of a bull and originated in the Middle East. Just as in Mesoamerica the people entreated a plumed serpent when asking for rain, Europe once had a serpent-bird goddess responsible for bringing it. But with the arrival of the bull and his rumbling thunder, the mother goddess became little more than the wife of the god of storm and rain. Astghik was the Armenian deity of love and fertility, but she was in charge of all things watery as well, while also being the wife of the god of thunder and lightning, Vahagn. The same happened with the Slavic Dodola and other similar figures; married off to male storm gods, they were none the less still in charge of the rain. Some retained the role despite certain adaptations.

Paparuda (in Romanian) or *perperuna* (in Slavic) is the rain dance performed by women covered in branches and grass in honour of Paparuda, the rain goddess in Romanian mythology, who lives on in the Balkans and corresponds with Dodola. In some parts of Albania and Romania, this rain ritual, held on the Thursday of the third week after Easter and also in summer when droughts are at their worst, is not the only one to have survived. The Frog Game is also still observed: a dance in which water is thrown over women and babies to attract rain while a child plays the part of a frog.

In Castrotierra, a hilly area in the Spanish province of León near the village of Astorga, they still have a rain goddess to whom they turn when the rains are late. She is a Romanesque image that appears in a procession, surrounded by banners, when a drought grows severe. Although the tradition appears to date back to pre-Christian times, this particular virgin has also been associated with a saint. They say that St Turibius, who was then bishop, grew angry with the people of Astorga during a drought and left, saying that he wished for "not even the dust" of that place to remain. He went away to the Holy Land, leaving seven years of drought in his wake. On his return he withdrew to a hill, taking with him a piece of wood said to be from the True Cross. The people of Astorga, knowing it wasn't just any souvenir that he had brought back, went to ask for his help. Turibius returned to the city he had spurned, and then the rain came. He asked the locals to go to Castrotierra (which comes from *Castro de la diosa Tierra* or "Fort of the Earth goddess"), where they would find a virgin who would help them ward off future droughts. Today the Virgin of Castro continues to receive their requests and, when "guardians of the land" decide that the time is right, they ask the bishopric for a special procession to Astorga to offer up a prayer for water.

Gimbutas found archaeological evidence that Europeans in ancient times often made "V"-shaped incisions as part of their rain rituals. It seems unlikely to be coincidental that the Taurus constellation isn't exactly in the shape of horns, but more like an upturned "V". What if all these beliefs have the same stars as their departure point?

In Greek mythology, the Pleiades and the Hyades were sisters, the daughters of Atlas. One day, the Pleiades went up to heaven in the form of stars and mounted Taurus for protection. There was a place reserved for their sisters alongside them. The Hyades' name come from the Greek verb "to rain". They are known as nymphs and rainmakers, sometimes also "the rainy ones". Their brother Hyas was killed by a lion while out hunting. His sisters started to weep, until they died of grief. Zeus, grateful to them for having raised his son Dionysus, turned them into stars which he placed over the head of Taurus; they formed the "V", the bull's horns. Aldebaran, the brightest, is Taurus's eye. Her appearance in the sky was then considered a precursor to rain, in reference to the tears shed for the death of her brother. But it was, even at that time, a very ancient story inspired by the stars themselves which had now become supporting actresses.

Nandeshwar and Nandini were about to say "I do" in the Indian village of Kalara, in Madhya Pradesh. Contrary to expectations, he was not wowed by his bride's dress—because she wasn't wearing one. All she had on was a yellow shawl, along with garlands of flowers and small, tinkling bells, and make-up so colourful it dazzled the guests. He also had a shawl on, as well as a sprig of fern on his head; otherwise he was naked, except for the all-over yellow body paint. Red dots had been painted around his eyes, recalling the image of the Hyades on Taurus's head. This wedding being out of the ordinary in so many ways, the bride and groom did not include their bank details on the invitation, their neighbours instead collecting all the money needed in advance. Some of those who turned up in their Sunday best still remembered the huge wedding of Ganga and Prakash, distant cousins of the bride and groom.

The wedding of Nandeshwar and Nandini, on the other hand, was more intimate, despite the fact that hundreds of guests attended, including about a hundred priests, a band and even a DJ to liven up proceedings.

The women, dressed in bright, beautiful saris, gathered around the bride, and the men the groom. A few small details you would usually expect at a wedding were missing. The bride and groom did not even promise to love one another till death did them part. Or not, at least, in human language. Because this was not a union between humans. Nandeshwar was a cow and Nandini a bull. They were married on a Monday afternoon in summer. It did not rain. It had not rained for a long time. Hence the wedding.

The point of the union was not to consolidate a couple's love, but to ask Indra for rain, as had been done in so many droughts in the past. In this part of India, weddings between frogs or bovids are nothing unusual, although not always are they intended to bring the rain; sometimes, as when the crops are flooded, they are also held to make it stop. The rain god here seems to have had the eccentric habit of waiting for humans to arrange marriages between frogs or oxen before deciding whether or not to send them a much-needed monsoon.

These singular Hindu wedding rites have been a frequent occurrence in recent years. Often the aim isn't only to summon or ward off the rain, but also to repel everything that going thirsty entails: poverty and hunger, as well as huge numbers of suicides in rural areas. There are an estimated twenty-eight daily suicides in India by agricultural workers. Official figures put the overall amount at 340,000 people in the last twenty years. This isn't primarily about too much or too little rain, but these are a factor. These people kill themselves after losing successive harvests while having bank debts hanging over them, or loan sharks after them, and energy bills they can't pay. People have taken one of two approaches in recent years: arranging weddings between cows or frogs, or challenging the government's authority.

Prayers for rain are also still a feature of the modern Catholic church; there is the *pro pluviam* prayer, but also *pro serenitate* to make it stop. St Isidore and St Mark are typically invoked in spring, but many villages have their own local virgin to whom they make offerings. The song dedicated to the Virgin of the Cave (Asturias) is the most well known, the Virgin of the Little House (Alaejos, Valladolid) also has her own: "Oh, Virgin of the Little House,/ you who have the power,/ unlock the clouds/ and let it rain!". The Virgin of the Mountain (Cáceres), the Virgin of the White Dove (Valencia) and the Virgin of the Holy Spring (Murcia and Teruel) are just some of the many to whom the thirsty turn. And then there is St Barbara, in whose name prayers and votive candles are offered to stop the rain. In La Montaña people first went out on a rogation after a drought in 1641. And it rained. As it always did. Was it truly a miracle or, as human experience suggests, is the moment of greatest despair always just before the end? In 2023, for example, votive offerings in Spanish villages were most frequent in May, increasing as the days and weeks went by. The rain, though it arrived late, did eventually come, at the beginning of June. Thus have propitiatory rituals always been retrospectively justified.

These Christian prayers are the equivalent to the rain dances and other propitiatory rituals that have existed all around the world, and are still performed in places as far apart as Mexico, Romania and Africa. The Cherokee nation have their own dance to bring the rain and good harvests, as was the case in ancient Egypt; dances were also about summoning the spirits of former leaders, which were enlisted against the bad spirits. The Zuni of New Mexico also have a special devotion to a rain god. *Viko lavi* is their festival of water and rain, and it is still celebrated in Mixteca Alta, Oaxaca. The central figures on that day are the *Ña tanjna*, men chosen to intercede between the earth and the rain-making gods. In Mexico, furthermore, though with regional variations, there are still the rituals of tiger fights and others involving weather magicians.

Islam, too, has its own invocation ritual, called *istisqâ*. This has occasionally taken place at the tomb of Yahya ibn Yahya, born in Algeciras and of Berber origin, known in his time as "the wise man of Al-Andalus". And it was to Algeciras, where the first mosque of Al-Andalus was built, that people went to pray for rain; there are written records of such gatherings as long ago as the thirteenth century. When the rain does eventually come, Muslims recommend staying outside and getting wet, "because that is what the prophet did". This year, in Spain, some of their *istisqâ* coincided with the prayers of Catholics. Separated by religion, united by a common cause.

Hindus and Buddhists, meanwhile, have their own prayers, in honour of Indra and Maheshwaranath. Kumari, the Living Goddess of Nepal, attends these rituals every year at the Indra Jatra festival in September. The difference here is that they do not pray for rain at the beginning of the rainy season, but at the end. The ritual begins with the raising of a column in honour of Indra, made out of a tree trunk from a nearby forest. Parades, masked dances and other celebrations take place over the course of a week in Kathmandu. The purpose of the event is not simply to ask for water; if they celebrate at the end of the rainy season it is because they are asking for the kind of rain that is good for the harvest; in other words, neither deluge nor just a brief drop.

Buddhism has good-weather monks too. In Japan they make cloth dolls to hang in the windows as a plea to the rain goddess, but these actually represent the shaven heads of Buddhist monks. And the request itself is different too: the *Teru teru bōzu* appear in people's windows on rainy days as a way of asking the heavens to go easy. Popular since the seventeenth century, they are hung up by children who repeat: "Good weather-maker, please make it good tomorrow". If they want more rain, they hang the dolls upside down, in which case they are called *sakasa bōzu*. But this tends to be the exception. Curiously, it seems that the Japanese have developed a kind of aversion to rain, since their mythology also includes Ameonna (雨女), a feminine *yōkai* spirit apparently inspired by a Chinese deity capable of summoning rain for the crops simply by licking her hand. In the morning, Ameonna takes the form of a cloud and during the night she turns into rain; she is said to be visible on rainy nights. Ameonna has a male equivalent called Ameotoko, and both terms are also used to describe the bad luck of people whom rain tends to follow.

Japan is not the only place where rain is seen as a bad omen. In Spain, rain on a wedding day has traditionally been considered a sign of an unhappy marriage: the bride will be crying for the rest of her life. This explains the tradition, dating back to the Middle Ages, of brides taking eggs to the Clarissan monasteries and asking for them to be offered up to St Clara for a dry wedding day. In recent years this tradition has gone further. Now anyone at all wishing to ward off rain on a particular day will take a dozen eggs to a Clarissan monastery.

Ever since human beings have known where rain comes from, stones everywhere have been marked, dances performed and songs sung, and people have worn outfits made of feathers and sprigs and bright colours, and cloud hooks have been used and rain stones like Ogotemmêli's too, all with the same end in mind: to ask the gods, amidst all their other chaotic goings-on, not to forget about the rain. From petroglyphs in the form of basins and channels, which, together with cave paintings, were perhaps the beginnings of rain magic, to *pro pluviam* prayers, Guerreran tiger fights and the *sakvari* chanted by Brahmin to summon the rain, rituals of propitiation have taken place all around the world. After grabbing power from the gods and spirits, human monarchs first of all became rain gods and then, as society became more complex, the task was delegated to priests, shamans and sorcerers. That was when the rainmakers appeared. Mortal men, they none the less had special powers, sometimes after being struck by lightning or because they had shared a womb with another master of water. In Spain and Latin America, these roles went on to be reversed, with men being made saints when they died and then, once in heaven, assisting the living from on high.

Not all of them became saints. Until quite recently, Spanish villages such as Cepeda and Cadaqués, and certain ones on Menorca, have had their own rainmakers, known across the Mediterranean as *trencador de les aigües*. The *trencador* of Cadaqués was particularly well known, going out on the Tuesday of the weeklong carnival there, stamping his feet, throwing himself to the ground and reciting, all the while flinging handfuls of soil in the air as a way of ensuring good rains for the year. There was a belief among African tribes that drought was a thing that brought war, and that a rainmaker, neutral and polyglot, would be the one to ensure peace. There is a curious connection between Zimbabwe and certain villages in Valladolid: while the women in that part of Africa dressed in black to make themselves look like the clouds and thereby summon them, during a drought the women of Valladolid would don widows' weeds as a way of invoking God's mercy.

If my grandfather shot at the clouds and my grandmother prayed to St Barbara, others shot arrows and rang bells to scare away the storms, or clasped the little fold-out Caravaca crosses that many Spaniards carry, or took shelter in exconjuratories. During antiquity people were already trying to deflect storms using "lightning stones", which were prehistoric hand-axes.

In one chapter in *Don Quixote*, the main characters follow the trail of some green grass to find a water source. But suddenly there comes a terrifying noise. Don Quixote's chest bursts with excitement as he perceives a possible new adventure, but Sancho, worried that it might be some beast of fable, immobilizes his steed Rocinante under cover of darkness by secretly tying up his legs in order to keep the knight from getting killed on another escapade.

The source of Sancho's terror is really the roar of a fulling mill. For if there is anything we fear as much as thirst, it is water, as attested by the number of origin myths that begin with a flood. In the 1950s, Dorothy Martin, a housewife from Illinois, claimed that Sananda, an extra-terrestrial, had come down to tell her that the end of the world was nigh, but that he could send a UFO to take her and anyone else who wanted to join them to safety. On this basis, Dorothy founded a sect called The Seekers, whose members believed that they were the elect, chosen to be saved from a flood that was coming to wipe out the planet. Hundreds of people sold their houses and on 21st December, 1954, they went out to wait for a spaceship bound for the planet Clarion. Among them were three undercover psychologists who wanted to see how these people would react when they realized that it wasn't in fact doomsday.

Sure enough, their pickup didn't arrive. And neither did the water. But, despite the cold, they didn't give up, and waited there until Christmas Eve. Although the prophecy was not fulfilled, they found an explanation; they said that Sananda had a very influential friend on earth who had interceded to stop the flood: Jesus Christ. Against all odds, and even though Dorothy Martin had to lock herself in her house and later flee the country, that failure strengthened the sect, and its founder became even more convinced that she had been chosen: it was thanks to her that the world had not been flooded. The sect grew. This contradictory response was due to what the psychologist infiltrators later termed cognitive dissonance. But there was something else: Dorothy was also dealing with one of the oldest and most universal fears: antlophobia. Even in the desert, where it rains the least, the fear of death by flooding remains acute. And it makes sense: when it rains on parched soil that is also bare of trees, it takes so long for the water to filter through that the effects can be catastrophic.

As well as in various creation myths, floods feature in the oldest written histories available to us. The *Epic of Gilgamesh* tells of the Great Flood a thousand years before the Bible. According to some of the peoples of Chile, we are descended from whales that were stranded on a mountain when the waters rose. For the San, water gave life but also took it away. This was reflected in their mythical stories, in which soaking a dead person in water could bring them back to life, and in which they had to look up at the point from which lightning issued to save themselves from dying in the rain.

But, as well as dances, songs and processions, there have traditionally been attempts to make it rain by sacrifices not always involving cattle or frogs: sometimes they were of humans. In the mid-fifteenth century, the Mexica and the Chimu sacrificed children and llamas in the Templo Mayor in Tenochtitlan, present-day Mexico, and in Huanchaco, on the north coast of Peru, with two opposite aims: to make it rain, and to make it stop raining. Archaeologists concluded that the former, accompanied by pitchers with carvings of the Aztec rain god Tláloc—"the one who makes things sprout"— may have been sacrificed to ask this god for rain. And right across from the Templo Mayor stood the temple dedicated to Ehécat, the Mexica god of the wind, who also made it rain. The drought of 1450–54 brought death to some, while others were compelled to sell themselves into bondage in exchange for corn. The famine was so extreme that it led to the sacrifice of 1454, the year "One Rabbit" in the Mayan calendar, a moment that possibly corresponds with the archaeological finds showing these rituals in Tenochtitlan and that is even depicted in the Codex Telleriano-Remensis.

For their part, the Chimu, who believed themselves descended from a being who came from the sea to found the city of Chan Chan, prayed to what was probably a sea deity to take the rain back up to the sky. To do this, children and llamas (brown llamas, specifically chosen when trying to contact the god of thunder) were taken in procession to the coast, where their chests were cut open, possibly to extract their hearts. The children were buried facing the sea, each with a corresponding llama looking back towards the mainland.

The Great Sprit began to dream. Fire and air came into being. And lastly, rain. A long battle gave rise to silt, to land and sea. The Great Spirit was pleased, and decided to stay. It asked the creator spirits to dream for it. As the dreams were woven together, fish, turtles, lizards, eagles, possums and kangaroos appeared in the world; then, when the kangaroo dreamed of music and laughter, humanity too gained the ability to dream.

In Australian aboriginal mythology, dreams and songs contain a creative power. The people believe that the rainbow snake engendered the primordial rain, but then stayed in the world, coming to reside in the Kakadu Falls. It is there that they go to invoke the rain. Among their creator spirits, those connected with the rain and clouds, the Wondjina, were indispensable. They first emerged from the clouds, created the sea and brought civilization to northern Australia. These were mouthless anthropomorphic beings who painted themselves on cave walls before returning to the clouds. One of them, Walagonda, became the Milky Way. There was a time when they did have mouths, but when they opened them the world was inundated with water. This is why Australian aborigines have always believed that a lack of a mouth wards off floods.

Extant cave paintings give pictorial backing to the myths; the Wondjina are depicted in reds, blacks and yellows. Were these paintings by their own hand, as the story goes? The timeframe of their execution is uncertain, but in accordance with some nearby fossilized wasp nests they have been dated to somewhere around 12,000 years ago. They lived in an area of Kimberley that was later flooded, forcing the population to flee when the Ice Age ended. This seems to be why their makers had such a focus on ceremony. Since their discovery, the paintings have been attributed to men, to owls and even to aliens. And the local people have assumed that, to avoid making these beings angry, and thereby avoid the floods and lightning they might otherwise send, their artworks must be cared for.

Among the Wondjina, Atain-Tjina stands out in particular, a rainmaker with his own myth among the Arrernte, a desert people of Central Australia. They say that he settled on the coast with other rainmakers whom he threw into the sea after a great betrayal. A gigantic water serpent swallowed them, but a few days later

Atain-Tjina went and asked it to vomit them up. Instead it gave them back transformed into smoke. One of them had plucked a scale from a fish and, after rubbing it against a rock, he himself became a cloud and from the sky he let his mane drop down. This was the rain.

Like so many Australian myths, this one takes place in Dreamtime. Thus Atain-Tjina eventually awoke, ascended to heaven, turned into a rainbow and joined the cloud-child. They set off together, heading west, and the rains came back. But then followed a long drought, which is the drought that Australia in fact suffers cyclically outside of Dreamtime and outside of myths.

8

Rainmaker

> The fact was that the clouds had that year withheld their moisture from the earth, and in all the villages of the district they were organising processions, rogations, and penances, imploring God to open the hands of his mercy and send the rain; and to this end the people of a village that was hard by were going in procession to a holy hermitage there was on one side of that valley.
>
> MIGUEL DE CERVANTES,
> *DON QUIXOTE*

> The people, down on their knees, implore St Isidore to send the rain that is so badly needed in these parched fields. Some of them take handfuls of the soil and kiss it and even, it seems to me, water it with their tears.
>
> ÁNGEL LERA DE ISLA

When my grandfather enlisted, my grandmother refused to kiss him goodbye. It was a slight he never forgot. When he got back, the villagers told him his betrothed was out in the fields, and he went to find her. There, on a patch of rugged land, a young man in uniform found a young woman sheltering a small hare in her arms. I was always intrigued as to what might have been said on this reunion, so I asked her one day. "Nothing," she shrugged. "What do you mean, nothing?" I said. "What was I going to say?" she said. And yet she remembers the scene with complete clarity, and has not forgotten how she grabbed and held on to his jacket when he left, what she had in her arms when he returned, and the contents of a poem he sent her while he was gone (possibly written by someone else on his behalf).

Those were difficult times for the land and the people of La Mancha. Drought returned to Spain in the 1940s, and it wasn't a brief visit. So severe were the effects that the patron saints were duly brought out for processions and rogations. My grandmother, a teenager at the time, went along with the older villagers. They carried the virgin shoulder-high so she could see how cracked and dry the land had become—in case her people's suffering had not been clear to her inside her chapel. They walked in procession along the Puebla del Príncipe road and, almost exactly where my grandmother would later go on to refuse to speak to my grandfather—also the place where I would later spend afternoons and evenings playing in the soil while my grandfather extracted water from a cave at the foot of Castillejo del Bonete—they turned and started back towards the village. The drought was bad enough, I suppose, to set off on a rogation, but not enough to walk all the way to the next village. Either that or it started to rain.

These two scenes, with their shared backdrop, are etched in my grandmother's memory, as is the song to St Isidore that is sung every 15th May when there has been no rain. In an act of pure anachronism, she sang it to me as a WhatsApp voice note. It is a ballad that, from the eighteenth century onwards, would have been handed out in chapbooks by itinerant blind men during processions and festivities. These blind men would go from village to village with a guide, singing and selling holy cards—which had something of the comic book about them—in exchange for a little money. There is still a saying in Aragón: "For the blind man to chant, it's money in advance." In those days, people remembered what they heard and passed it on orally. It might not seem like something that would have been destined to last, but chapbook literature had a revival in rural Spain and continued to have a presence there midway into the twentieth century, when my grandmother was a young woman and drought was once again wracking the land. It is therefore not surprising that different versions of "St Isidore the Farmer" exist across La Mancha and Jaén as well:

> *When St Isidore the Farmer*
> *went out to work the land*
> *it was late on in the day.*
>
> *The other farmers*
> *all looked on in envy*
> *when they saw how well he did*—hey.

The story recounts Isidore's miraculous life, with all the over-egging characteristic of this kind of oral literature, and all the emphasis on the greatness of small deeds. Then my grandmother came to the part about thirst with which the song concludes:

> *What do you want, my master,*
> *what should I say?*
> *That at the top of that rock*
> *springs water clear as day?*

Girls and boys of my generation no longer memorized chapbook ballads, but we had our own song on the same subject. It began: "Let it rain, let it rain,/ Virgin of the Cave." And, in a moment of clear childish egoism, it concluded with: "Let the windows at the train station explode,/ but not my own." Since there was no train station in our village, I pictured the windows at my school exploding. It was like the Móstoles girl who, asked by a journalist what her dream was, said her greatest desire was for her school to be destroyed. "Let it rain," we sang, but I know we also sang this once rain had begun to fall and we were splashing about in the puddles and doubtless, too, after my grandmother had said her rain-prayers and was already deep in the making of our rainy-day pease pudding.

On the same day my grandmother sent me this song about her saint rainmaker, I happened to see a film called *The Boy Who Harnessed the Wind*. William Kamkwamba, though he wants an education, has to stop going to school when an extended drought ruins the harvest and his family is left destitute. While others in the village are driven to despair, William manages to access his local library, where he finds a book that lights up his imagination. If he builds a windmill, it says, he can use the wind to activate a pump that will bring up the well water, meaning they can irrigate the village again. No one he speaks to offers any encouragement, but he never gives up. All he needs is an object from his father, but that object happens to be all they have left since being forced to sell the roof of the house in order to eat once a day.

I won't apologize for the spoilers this time, because William Kamkwamba's story is true and recent. He has written about it and told the story himself in talks and in the book that inspired the film. It happened in Malawi in 2002. Although it doesn't feature in the book, there is a moment when William's mother, almost as if she had been listening to my grandmother speak, says: "Although they would have prayed for rain, the reason our ancestors survived was that they were united." And that makes me wonder if my failure to understand that my grandmother's incessant prayers over the years conceal more than just words and faith has been arrogant.

She prays every day. For her daughter in her work, for her son-in-law who works on the roads, for her grandson who is unemployed, for the granddaughter who is away travelling, for the neighbour who is hoping for rain, for the other neighbour who is ill and has just become a widow. My grandmother praying for all of us is more than a religious experience; it is an old woman dedicating her time to looking after others without physically touching them. If she prays now for St Isidore to bring rain to the fields, she does it for the same reason my grandfather cleared the tracks, even though he himself didn't have a car. She isn't in danger of a harvest being spoiled, save that of her small number of olive trees, the oil from which we as a family consume and we already know full well is petering out. She has also asked my mother to do the processing now, saying that "a dead person covered in stains doesn't look nice". She has always been obsessed with her own death, but now she talks about herself as if she is already dead; any time anyone leaves the house, she asks them to make sure her shroud is ready. So she prays for rain like somebody planting an olive tree for their grandchildren, holding no hope for herself other than being presentable on her deathbed, make-up done, clothes ironed.

Praying and clutching a rosary will not make the rain come, but it will make you feel part of something bigger than yourself. When you pray, you meditate and at the same time wish goodness upon others, which strengthens mental bonds; this after all is the true meaning of the rosary, which was lost when Christians made an incorrect translation of the Saracen rosary. As Mario Satz writes in *Pequeños paraísos* (Small Paradises): "In fact, the rosary, or *mala* as it is in Sanskrit, comes from India and as far back as the *Bhagavad Gita* its thread was referred to as the soul or *atma* that weaves together all worlds and all living beings." The root of the term "religion" is *re-ligare*, or "to reunite". Perhaps the greatest paradox of humanity is that what began as a means for unity went on to become a weapon of social control and now divides us. Before it was a private and individual matter, before becoming the opiate of the people, religion was our glue. That is why going on a pilgrimage is about much more than believing that the saint will really be able to mediate between heaven and earth: it's a way of sharing and warding off fears, and it is just as valid as any rain dance.

In one sentence, William's mother essentially summarized what my grandmother has been gesturing towards of late, which is also what I wanted to get across in this book: that the current, terrifying drought has always been with us, and has never destroyed the world or humanity. But it did lead to the collapse of civilizations in which walls were built, in which water was administered unjustly, with the greedy few condemning the masses to destitution. Those who adapted to the changes in their environment managed to hold out, as did those who spoke those first few words: *you, I, us, give, flow*. With those words they built stories. It matters little whether they were religious beliefs, myths or legends; what matters is that they were shared. And we still have time to learn from them. Only by recovering social cohesion and nurturing an awareness of our common humanity will we be able to move forward or, at least, soften what now seems to be the inevitable blow to come.

※

Isidore was known as an idler. Always late for work, people said, the last to pick up a hoe, and with a tendency be absent at the most inopportune moments. His excuses always had to do with spiritual enrichment, and he had no qualms admitting that he would sooner take time off than risk skipping prayers or missing Mass. Last to start and first to finish, and during his brief labours he would always find time to stay on good terms with his creator. He was once asked who helped him, to which he answered, God himself. Rumours started to circulate about his oxen ploughing the soil all by themselves, about the fact he could conjure water with nothing but a word, and indeed make it spring forth by striking the ground with a stone. This was in twelfth-century Madrid—then still known by its Arabic name of Mayrit, although it had been conquered by the Christian Alfonso VI in 1083. Isidore was one of many Mozarabs who had arrived with their families to repopulate the area. The line between envy and admiration is a very blurry one. Isidore provoked both in equal measure, and it was this that led people to denounce him.

Alerted by the other labourers, Iván Vargas, owner of the land Isidore worked, had been spying on Isidore one day and wished to reprimand him. But then, the story goes, he grew thirsty, and Isidore thrust his staff into a stone and made water appear. Curiously, this was a story that had already been told of Neptune, Moses and St Turibius. When the farmer's life came to be recounted, it was angels pulling his oxen. And both his laziness and his ability to command water featured. His son was said to have fallen down a well, with Isidore then going and asking the water to rise in order to bring the boy up. And the son was saved.

His connection with water won him fame as a diviner and rainmaker. Such was the animosity aroused by the farmer that Lope de Vega's *St Isidore the Farmer* features Envy as a character. He enters the scene preceding the marriage of Isidore and Maria, whose beauty the farmers compare to snow in December, and he cries out:

> *A farmer I envy*
> *for he presumes to make off*
> *with the states that I lost in the war [...]*
> *All the poison locked up*
> *in my burning chest must out,*
> *and Isidore die, die!*

The story goes, however, that he lived to ninety years old, even though we have recently learned that he would have been unlikely to make it to forty. Although he attained a certain renown in life, his burial was such a humble affair that he didn't even get a gravestone, or anything with his name on it suggesting that this was where he lay, interred in the church where he so often went to escape his work. Or so the story goes. It was wet in that particular spot, and water flowed beneath his body. His grave, the story goes, flooded numerous times.

Isidore's life unfolded during the Medieval Climate Optimum, a kind of 300-year spring—with a few hiatuses—bookended by centuries of drought. Between approximately 900 AD and 1200 AD, temperatures in the Iberian peninsula remained stable, characterized by mild summers and winters, and rain that almost always arrived in good time and in good amounts. Frosts were a rarity in May, summers stayed dry and warm, but by spring rain had already fallen in reasonable quantities. After more than two centuries of good harvests, and in spite of a number of wars, the population of Europe tripled, while cathedrals were springing up everywhere, along with stone bridges spanning all the newly swollen rivers. It was also a time of great pilgrimages. To the people it seemed that God was pleased. But it was up to them to keep it that way, whether via offerings, pilgrimages or new cathedrals. The Vikings were also expanding across the northern seas and reaching hitherto ice-locked places, such as Greenland and Iceland. Spain became a great exporter of wool and Alfonso X the Wise created the Mesta, an association of sheep farmers in the central plain, the legacy of which can be seen in the royal drovers' roads. In Europe, crops grew more widely, with the plants themselves growing taller too. The absence of extreme cold favoured their growth, as did the founding of villages and towns in mountain areas that adapted to agriculture and livestock by cutting down many of the forests. In that long spring, droughts were scarce, but there were exceptions that devastated some areas of Spain at specific times, as happened in Galicia the year Isidore died.

In 1212, as the dry conditions in the Meseta Sur intensified and ice was creeping out across northern parts of Europe, Isidore's body was exhumed. To the surprise of all, his body had not rotted. There was still skin on his bones, and they even detected a certain flexibility to his neck. Not just that, but he was said to give off a pleasant odour. There are accounts of Alfonso VIII returning victorious from the Battle of Las Navas de Tolosa and wishing to thank a certain country peasant who had shown him a hidden path that enabled victory over the Almohad caliph; seeing the exhumed body, Alfonso said it belonged to the person who had appeared to him, and he ordered the body to be placed inside a gilded coffin. The body began to generate as much interest as the water-related miracles attributed to Isidore both in life and death. The sight of his uncorrupted body led to his being unanimously beatified, and with that grew the belief that he could go on summoning the rain even in death. And this was when his story began to be rewritten. No longer was he an idle farmer envied by all and spied on by the landowner, but rather a being of light who made water spring from a rock by striking it with his staff. How could anyone have reprimanded a saint, if the things he did had already been foretold in ancient myths? In this new version, dating from the fifteenth century, Vargas dashed to tell his wife about the tenant farmer who was actually a saint, and could get his oxen to plough the land unaided. The angels were yet to join the action.

If there was nothing separating him from heaven, communication would be easier: the decision was taken to open the coffin, to make for less interference, leaving Isidore's intercourse with the rain unobstructed. But this then seemed insufficient to people, and they decided to take him out as part of the procession, so he could see the parched fields. The severe, countrywide drought of 1231 was only a sign of things to come, and the Madrileños of the time duly paraded the body of the one they believed to be a rainmaker. And the rain soon came back. It had worked, they said. So followed a series of post-mortem exhibitions and excursions that went on for 900 years, and reached their peak during the centuries of the Little Ice Age.

The cold arrived in Greenland, Iceland and the Arctic a little after 1200 AD, carrying on down through Poland and across the Russian steppes. It was another century before its effects were felt in Europe. Then, from the fourteenth century onwards and for more than 500 years, what has been called the Little Ice Age brought in not only ice, but devastating storms, cold summers, severe droughts and gale force winds. The weather was constantly changing. During some winters, rivers such as the Thames froze to a depth that made ice skating possible, and balls and fairs were held on them. During a succession of winters, the Ebro could be crossed in places on foot. It was in this context that a new pictorial genre emerged, in which Pieter Brueghel the Elder was a leading light, as well as a literary genre: horror. Summer sometimes simply didn't come.

The causes of these sudden, unpredictable changes are still not known for certain. From the seventeenth century onwards the effects were global, and these persisted to greater and lesser degrees until midway into the nineteenth century. But the most widely accepted hypotheses point to shifts in the rotation of the Earth's axis and periodic changes in solar activity. These theories, however, came later. At the time, most people just thought their god was angry.

In 1315 it started to rain in northern Europe and didn't stop for several months. So much water destroyed the crops. The same happened in 1316, but this time the bad harvest had to be added to the previous one; these years marked the lowest levels of cereal production in the whole of the Middle Ages. Country folk, increasingly debilitated, had lost harvests and animals, and could no longer afford the price of bread. People went from town to town begging alms, and eventually found their way to the cities. In Monty Python's *Knights of the Round Table* a man goes around piling up a cart with the dead. He is no fiction.

In 1317, when famine struck, livestock were affected by disease, which resulted in depleted flocks and in turn less fertilizer—this, on a predominantly agrarian continent which, with a few exceptions, revolved around a cereal-based subsistence economy. It was now that pork became such a vital component in people's diets. The summer of 1317 was just as wet as the previous ones, and special processions were organized to try to get God to turn off the tap. The assumption was that this was all a punishment after the plenty of preceding times, when almost half the woodland in Europe had also been decimated to create space for arable farming. Until then, common land had allowed the poorest country folk to survive. But they began to be deprived of those rights too.

In 1322 the North Atlantic Oscillation (NAO) index was reversed. When the index is high it makes for drought in southern Europe, while low means intense cold. These cycles tend to last around seven years. Perhaps this is why the drought in the *Epic of Gilgamesh* devastates Uruk for seven years and why the Famine Stela speaks of a seven-year drought in Egypt. When the NAO index swung back again, people's despair was over. Briefly. What then followed was a seventeen-year drought. The abrupt changes over the course of that century had merely been a sign of what was to come.

As if that were not enough, the beginning of this period is believed to have coincided with Mongolians departing the Gobi Desert to escape thirst at a time in the fourteenth century when Europe was inclement and Asia parched. It seems they were unwittingly accompanied by rats infected with *Yersinia pestis*. Although the precise origin of the Black Death is still unknown, a pair of scientists came across a cemetery in Kyrgyzstan where an unusually high number of burials had taken place between 1338 and 1339. Inscriptions on the gravestones pointed to the very thing later borne out by analysis: "Pestilence". After analysing the bones and finding traces of *Yersinia pestis*, the origin of the Black Death was moved forward a decade and its location altered. This study is of interest to us here because it concluded that the drought would have suddenly reduced the capacity of that ecosystem to support rodents, which would have forced their fleas to look for other hosts in camels, sheep and the herders themselves. When the plague arrived in Europe almost a decade later it found the ideal conditions among a population on its knees after drought, flood and famine. The already weakened population put up little resistance. The first plague epidemic broke out in 1351. It then recurred every ten years.

On 16th January 1362, the waters of the North Sea covered a large part of the British Isles, the Netherlands, Denmark and northern Germany in what is known in Danish as *Den Store Manddrukning* and in Dutch as *Grote Mandrenke*. The meaning is the same: "the great drowning of men" (it was also called the Second Flood of St Marcellus). It resulted in the deaths of between 40,000 and 100,000 people, the destruction of towns and cities and the submergence of the low-lying city of Rungholt, a kind of Germanic Atlantis shrouded in so many mysteries and legends that for a long time it was believed to have existed only in fable. The bells there are still said to toll beneath the waves.

Excessive rainfall and the Hundred Years' War worked together to the same end, aggravating the generalized starvation during the first half of the 15th century. Between 1433 and 1438, hunger turned to famine. Following this, a fall in the price of grain drove many arable farmers to turn to livestock instead.

Three centuries after the *Grote Mandrenke*, the London baker Thomas Farryner took one last look at the fire in his hearth before going to bed. Everything appeared to be as it should be, with hardly any embers remaining, and he stirred around in the brick-paved fireplace to extinguish them. A few hours later, Jane, Samuel Pepys's maid, went to wake her masters: the heart of London was burning. People threw themselves in the river, while others sat in their homes just waiting for what seemed unavoidable. Even so, when the mayor received the news, he played it down. "Pish, a woman might piss it out," he is reputed to have said. But the drought of the previous year had left the city of thatched roofs and narrow streets tinder-dry, and high winds that night ensured that it burned down almost entirely between the 2nd and the 4th of September 1666. The drought and the wind were compounded by the mayor's initial contempt and subsequent hesitation, as well as the search for culprits among the residents. Foreigners, mainly French and Dutch, were accused and attacked. Another obvious culprit was identified: a woman running along with a laden apron. People assumed she was an arsonist and that she was carrying explosives. But she was no more than what she appeared to be: a woman running away with food, and the alleged incendiaries just chickens. All this allowed the fire to advance and engulf the homes of some 80,000 people. The death toll is unknown because only seven people were registered. The poor didn't count.

Farmers did not usually become saints. Neither did married men, and it was even less usual for someone's wife to do so, or indeed for Christians and Muslims to share a saint. But then again, much that was happening at the time was far from normal. St Isidore, and all the stories associated with him, imbibed many of the miracles associated with water-commanding saints from desert regions. In the Iberia of that time, it was enough for a group of people to consider somebody a saint for them to become one. The corresponding bishop would accept the nomination, sainthood not yet being exclusively the pope's to confer. Isidore's religious syncretism had echoes at other levels, given the way he also united religious and lay people, city and countryside, nobility and commoners, married and single, pork-lovers and pork-haters. He was a peaceable saint who drew factions together, and his renown went on growing over the following centuries. Today, as the main character in so many rain-summoning rituals, he has transcended religion altogether.

The public display of his cadaver became increasingly theatrical and morbid in a country that had not yet expelled death from its towns and cities, or even from its homes. Pulling out a lock of his hair was thought to be a way of conjuring the rain, but another idea then arose, of his being able to cure kings, and with it the tradition of placing Isidore in the royal bed. The story goes that when he was first brought before a Spanish king, he was missing three toes. A tooth as well, apparently. Such was the fervour aroused by the dead man that Carlos II's locksmith had pulled it out and kept it for himself, for whatever reason. The king found out, and the locksmith had no choice but to hand over the tooth. As if hoping for some sort of ghastly tooth fairy, the king sewed the tooth into his pillow. Queen Juana, Enrique II's wife, broke off his arm. One of Isabel la Católica's ladies pretended to kiss his feet, but in fact she was taking a big mouthful; she bit off his big toe. But she was unlucky: her actions were said to have caused the Manzanares River to burst its banks, meaning Isidore's progress was halted. The woman saw what she had done, confessed, and the waters duly receded. Supposedly, a salve was even made from one of Isidore's fingers. And when it came to Carlos III, the mere presence of the farmer's corpse was not enough; he also had the skull and shinbones of Isidore's wife brought to him, proceeding to kiss them with such devotion and tenderness that she has been known as St Maria of the Head ever since. She is often depicted with a jug of water. These accounts are so similar that they scarcely seem true, putting one in mind of cautionary tales; perhaps it was the kings themselves who had an interest in their dissemination.

Only one king refused to share a bed with the saint. Felipe IV's response, when it was suggested Isidore be brought to the royal bedroom, perhaps implies that the cadaver did not smell quite so much of roses as all the stories said. If the saint was capable of miracles, said the king, surely he could perform them remotely.

Thirst, allied with the cult of St Isidore, was a key factor in Madrid's becoming the Spanish capital. Since the time of the Catholic Kings, the court had been moving ever closer to Madrid, in large part because of the water sources there, but especially because of their curative properties, which were strongly linked with Isidore. Located over an aquifer, Madrid was supposedly full of miraculous springs. The Catholic Kings began increasing the number and duration of their visits, and later monarchs followed suit. One day, the young prince Felipe and his father, Carlos V, both came down with a fever. The boy went and drank from Isidore's miraculous well (located where he was said to have called forth the water), and the saint was credited with saving his life. When he went on to become King Felipe II, he relocated the court from Valladolid to Madrid, which became the capital in 1561. He felt himself to be in Isidore's debt, and so moved heaven and earth to have him canonized, which was also a way of ensuring that his court would be at the centre of the Hispanic world. Felipe II was unusually obsessed with the bodies of the saints, and not only Isidore's; he had the palace at El Escorial littered with their bones as a way of protecting it from storms. On his deathbed, he turned to relics in an attempt to cling to life. He thought that the skulls of various saints, together with the jawbone of St Agnes, the knee of St Sebastian, an arm each from St Ambrose and St Vincente Ferrer and the rib of a bishop would be enough to stave off the death that was calling to him. They weren't.

Various churches outside of Madrid have claimed to be in possession of relics of St Isidore. In Argentina they didn't want to be left out either. On 12th October 1928, the same day that Hipólito Yrigoyen was sworn in as Argentine president for the second time, someone in a nearby town wrote a curious letter to the king; the people of a place called San Isidro were requesting a piece of their namesake saint. An entry in the blog of the National Library of Spain, which is based on its newspaper archive, tells of the town, not far from Buenos Aires, having been founded by a Spanish captain who also installed a chapel dedicated to the saint. The inhabitants had previously asked for a souvenir, but the urn they had received required ten keys to open, and those keys were held by the King of Spain. So one day Alfonso XIII received a strange letter signed by Ramiro de Maeztu, then Spanish ambassador to Argentina, applying to him, the bishop and the mayor, for a piece of St Isidore's body.

After long deliberations, they decided to agree. The coffin was opened and a doctor said a prayer, scalpel in hand, before removing a piece of tibia. The Argentine town was decked out to welcome it. And so the body of St Isidore began its post-mortem transatlantic adventure at a time when the farmers of drought-stricken Spain could rarely afford to travel as far as the sea.

Having become the central figure in rain rituals during times of drought, canonized in 1622 and often depicted together with oxen, St Isidore became the successor to the celestial bull and the rain god of those ancestors who drew animals in the sky, linking the stars together as a way of establishing whom to ask for water. Not for nothing is 15th May his day, given that this is when the sun shines on Taurus, and bulls are still sacrificed in his honour in Madrid.

Devotion to St Isidore began only at the end of the thirteenth century. His posthumous miracles, performed by invocation, have always had to do with the healing of the sick, strengthening of faith and, especially, petitions for rain. A decade after the 1231 display, the body was again exhumed for the same purpose. In 1261, there was also a *pro pluviam* rogation. In 1272, it was taken in procession to the Basilica of Our Lady of Atocha. In 1426, during another drought, the body was once again carried through the streets. Until then, it had been an object of veneration for monarchs and landowners only, not the peasantry. When the famous spring appeared in the mid-fifteenth century, the worship of the saint spread to the common people, and this was when pilgrimages began. These initially consisted of going to drink from the spring as a cure for fever and, at the same time, going inside the chapel to ask the saint for rain every 15th May. In 1709, the image of St María of the Head was added to the rain procession. Both were transferred to the parish of St María. The 1780 *pro pluviam* rogation took things further still. Not only did Isidore's wife accompany him; they were both placed in the convent of the sacramental nuns with the idea that, until it rained, there they would stay. Eleven days later they were finally able to go home.

There was an added importance when Isidore's body was exposed for veneration in 1896, with two things being asked of him: rain, but also an end to the war with Cuba. On 4th May it began to rain during the procession, and the people there took this as a miracle on the part of the farmer saint, with the newspapers reporting as much the following day. It rained across almost the whole of Spain with no let-up for several days. (We will return to this specific rogation because it was particularly important in a more recent episode of our search for water.) When the rain came on 15th May, the body was once again exposed, and by now it had achieved such fame that 300,000 people went to view it. There had been other displays for different reasons (kings falling ill, invasions by the French, anniversaries), but in the twentieth century no further processions took place to ask St Isidore for rain until 1947, which was when my grandmother learned the song.

The dates of these displays and petitions were not random, nor were variations in the way they were performed. The majority took place between the fourteenth and late nineteenth centuries and as we have already seen, though the dates and causes of the Little Ice Age are still up for debate, we know that it lasted for at least five centuries. *Pro pluviam* rogations vary according to the severity of the drought, so church records give us an insight into the worst droughts in Spain over several centuries. The rogations vary by degree: minor (a simple prayer), medium (display of the intercessor), serious (masses and processions with the intercessor), major (procession with the intercessor outside the church) and extraordinary (pilgrimage to another place of worship). This tells us that the droughts of 1709 and 1780 were especially devastating, at least in central Spain. The fear of drought and its consequences—that is to say, crop failure and famine—saw St Isidore's popularity soar, his remains carried far and wide, and his being made patron saint of farmers in 1960.

Whatever the causes, from the beginning of the fourteenth century when the northern hemisphere began to cool, driving the Vikings out of Greenland and triggering the floods that plunged Europe into famine between 1315 and 1317, things were never the same again. The situation worsened and reached a peak between the seventeenth and eighteenth centuries, before stabilizing around the middle of the nineteenth century. By that time it was a worldwide phenomenon. Darwin wrote in a notebook about the drought, or "Gran Seca" that devastated Argentina between 1827 and 1832, when it became difficult to use the river because of the stench: "hundred[s] of thousands carcases dead on banks (fall down barrancas) float in water: could not pass many of the streams for smell—it would be said some great flood had killed all, especially as after it all rivers were very much flooded corresponding deposit."

It seems unfathomable that a world wholly dependent on rain had no systematic way of recording climate conditions and relied solely on people's memories. But something had begun to change: the first documented mention of *cabañuelas*—a forecast method for the twelve months of the year based on the weather in the first twelve days of January—dates from this period.

There are various explanations as to why the cult of St Isidore grew in later centuries and why other climatic intercessors emerged or became more popular, such as the Virgin of the Cave, St Barbara, St Mark and figures equivalent to St Isidore in other European countries, such as St Medard in France, who in the fifteenth century became the protector of the French royal family. However, veneration of Isidore appears to predate these others, and there are records from earlier centuries of a tale about a violent storm coming along when he was a child and an eagle sheltering him under its wings. It is also a long-standing custom that rain on his feast day, 8th June, is a sign of forty more days of rain, as is also the case with the English St Swithun of Winchester, who was being venerated as long ago as the tenth century and whose feast day, 15th July, is taken to herald forty days of either rain or drought. The increase in *pro pluviam* rogations in the early seventeenth century even makes an appearance in *Don Quixote*: meeting a group of penitents bearing a statue of the Virgin Mary as a rain charm, the knight errant takes her for a kidnapped damsel and tries to rescue her. He ends up just as soundly beaten as in so many of his other flights of fancy.

Cervantes experienced not only the thirst that came with the driest moment of the Little Ice Age. If he is believed to have died of diabetes, a disease not yet diagnosed at the time, it is because in addition to being constantly tired, he was always thirsty. A thirst that not even water could quench. What better description of the polydipsia he suffered than his own words: "the entire sea could fit inside it [his belly]". Physicians prescribed him country air and wine from La Mancha, specifically from Ciudad Real, to combat his constant thirst—in a sense making them great publicists, given all the vineyards that then began to spring up in the countryside around the place of my birth. They say that in 1616, the year Cervantes died, springs in the arid regions of Spain ran dry and rogations became commonplace, yet the crops still failed.

Our ancestors could have overcome the loss of one year's crops, but two in a row would have been extremely challenging. In a country where agriculture was still just about subsistence and irrigation techniques had not yet come in, people responded in two ways: through religious faith, and political rebellion. The former was intended as a preventative, with rogations used to call on the rain; the latter was a reaction when widespread thirst and abuses of power converged.

The religious response may also have been an attempt to ward off rebellion, given that the rogations were not spontaneous. Although traditionally farmers requested them, they had to follow a process that meant applying to the respective bishop, who would be the one to decide the date. If they failed to follow this protocol, people taking part in rogations were liable to be excommunicated, as happened on a number of occasions in the seventeenth century. Rogations were, finally, far more than a hopeless petition: they were also a way of managing and avoiding rebellion among a thirsty, starving and discontented population at a time when droughts were growing ever longer and more severe.

Both of these responses to thirst tended to come about in the spring, and May in particular. There has been some research into the link between warm weather in late spring and the outbreak of revolutions and wars. And links have indeed been found, so that the Spanish saying "spring alters the blood" may indeed have a basis in fact. But there may be something else at play: rogations and uprisings have often happened at times when the price of bread has peaked, with reserves from the previous year's grain harvest running out. This is also the last chance for any rain that has not yet come, which might in turn save the coming harvest.

It is a time of year when it is easier to gather together, whether that be in faith or an attempt to point the finger. Few things have had such a unifying effect as people's patron saints, their rainmakers or rulers, or indeed neighbours who seemed too free, and who knew the secrets of plants and births. In the same way, when people have been most in need of rain, they have joined together for rogations, riots or sacrifices. In that order, very often.

Although there was no single reason for all of these events, the fact that they coincided, often after several failed harvests, seems to indicate that there must have been a common factor, one that transcended peoples, beliefs, nations, empires, continents and indeed the whole of humanity.

I was born on the anniversary of St Isidore's canonization. For some reason—perhaps because I hail from a family with strong ties to the land, with at least seven of the last nine generations day labourers—I have always felt a certain familiarity with and fascination for the man I imagine in the way he is represented in our village: long chestnut-brown hair, goatee on his kindly face, leaning on a hoe. A reconstruction was recently made of his face, and he did look more or less like this. I find it comforting somehow that a farmer who was famous for being idle and kind should still have so many admirers almost a thousand years after his death. And I also enjoy the idea of him skipping work to feed his soul, with merely a word to his master to say that he needed to go and meditate and that the angels would cover for him.

While Madrileños continue to drink from Isidore's fountain every 15th May and to ask him for rain, the day is still celebrated in some villages. "St Isidore the farmer," they say, ever hopeful, "give us rain and take away the sun." His adventures have continued, and in 2022, on the 400th anniversary of his beatification, his body was exhumed and displayed once more. In 2023, after drinking water from his fountain, I heard a priest praying for rain in his parish while the congregation (quick to forget the pandemic) went on kissing a relic.

There is a chapel in honour of Isidore on a hill in Terrinches. I spent every 15th May of my childhood there. I have a photo of me grinning happily, sitting astride my grandfather's moped. My brother is with me on the motorbike and we are wearing the same tracksuit or pyjamas. My grandfather watches happily over us, while everyone else in the background has a mouth stuffed full of food, that day really being about eating, drinking and dancing. I have other photos of my brother and me among the craggy rocks. I don't remember much about the festival, apart from that feeling of shared joy, the food and the rocks. That's what the photos really show. The saint was nothing more than an excuse to get together, in the hope that he would grant us something we all needed. But we children didn't know that and, moreover, the saint was eclipsed by another guardian of water in the religion I share with my brother, which is none other than childhood itself.

Soon after my grandfather's death, something of his came into my hands. It was a Y-shaped branch. An unfinished catapult, I thought. I don't like the idea of killing birds, but I do like that this thing was left incomplete, in that this encapsulates death: all the things we leave unfinished, and that other people wish we had finished, define the parting. A catapult, a book, a look, an answer. After a number of years of keeping this supposed catapult with me, I had to get a good way through this book before looking at it with new eyes: had I been deceived, did it merely *look like* a catapult? The only person who could answer that now was my grandmother. But it was late at night. Only afterwards did I learn that she hadn't slept that night, and that I had made a mistake which smacks so much of the living: for her, the next morning didn't arrive. For all that she talked about the day of her death, and left clues, for all that she was constantly saying Resurrection Sunday was "a lovely day to die", I fell for that more than mortal mistake: deferring a question for a day that might never come. I had to put lipstick on a mouth that would never speak to me again—a task she had requested of me for years. I was left with an object that might have been an unfinished catapult, but by the same token could have been a water diviner's stick. And I was left with a doubt. My grandfather's Y-shaped stick came down to me, as did my great-grandfather's flint for lighting fires, and I have kept them together ever since. Perhaps what I have been carrying around is a kit for invoking storms and detecting water under the ground. Or perhaps not. There is nobody now who can tell me for certain.

9

Drought and disorder

Then it came, the happy moment,/ when the tortilla was flipped:/ the poor got bread to eat/ and the rich could eat shit.

FOLK SONG OF VALLADOLID WOMEN WHO
SPARKED THE 1856 FOOD RIOTS

Three things exercize a constant influence over the minds of men: climate, government and religion.

VOLTAIRE

On 14th July 1518, in Strasbourg, capital of the Holy Roman Empire, Frau Troffea started dancing in the middle of the street. Not so strange in itself, if it weren't for the lack of music. Then several men joined in with her frenetic jig. After dancing non-stop for hours, she fell to the ground in exhaustion, gave a shudder, and fell asleep. The more recent arrivals went on prancing and cavorting, and eventually she herself ended up joining the dance once more. The hours turned into days, and the days into weeks. A month later, there were around 400 people there, still dancing, completely unable to stop. They cried out for help, because their fizzing bodies wouldn't let up, not even when they wanted to eat, drink or sleep.

Feet bleeding, some fell down in pure exhaustion, while others suffered haemorrhages and heart attacks.

The priests thought they were being spurred by a divine punishment on the part of St Vitus, while the doctors put it down to a heat in their blood so severe that not even bloodletting would cure them. The former came to the conclusion that they should be sent on a pilgrimage to repent for their sins before St Vitus, but the doctors said the only answer was the very thing that had driven them to do what they had done in the first place; they should dance till they dropped. But that didn't help. At a certain point, that suicidal dance did come to a sudden end. The survivors were told to go and commend themselves to St Vitus as a way of purging their sins, but what had possessed them remained a mystery, as did their being unable to stop themselves even when they realized death was waiting in the wings, holding invisible instruments.

Nearly a decade later, the doctor, alchemist and astrologer Phillippus Aureolus Theophrastus Bombastus von Hohenheim arrived in the city—fortunately for his new neighbours better known as Paracelsus. He was amazed to hear of sinners whose divine punishment comprised this uncontainable urge to dance. Looking into the matter, he came to the conclusion that Frau Troffea was to blame; her dancing, in his opinion, was nothing more than an attempt to anger her husband. Some time later, he claimed that "prostitutes and low women" had been the victims of a phenomenon he called *chorea lasciva*, which he defined as a kind of predominantly female hysteria affecting free, lascivious and disrespectful individuals he called "choreamaniacs". Furthermore, he said that they deserved to be locked up in the dark, with only bread and water. He also suggested burning wax or resin effigies of them.

But this was neither a Renaissance rave nor an isolated phenomenon. It also wasn't the first manifestation of a dancing mania, popularly known as "St Vitus's dance", to afflict the inhabitants of a European city. In Alsace itself the same thing had happened just eight decades earlier and in 1020 the residents of Cölbigk, Saxony-Anhalt were gripped by it too. After outbreaks in south Wales and in Erfurt, Thuringia, in 1278 the people of Maastricht continued bobbing away—literally now—when the bridge they were on collapsed under their weight; on they danced, utterly unconcerned for their lives, as the river swept them away. As in Strasbourg, the survivors were unable to explain why they had begun dancing or why they couldn't stop. There were similar and ever more frequent outbreaks in other central European towns, especially from the fourteenth century onwards, and it wasn't until the end of the eighteenth century that they began to tail off. The last one occurred in the mid-nineteenth century.

Although these "dance epidemics" were looked on for centuries as an expression of the same phenomenon, the reason behind them is still unknown. For contemporaries of Paracelsus, they were the result of demonic possession. But, over time, researchers began to suggest that they could have been cases of collective hysteria unleashed by the constant stress of floods and droughts. Some theories point to the effects of people going hungry, this itself the result of climate change. Or people might have eaten mouldy bread contaminated with ergot, a fungus that can affect cereals such as wheat and barley and has an effect similar to LSD. And yet we would still return to the climate, ergot being particularly common in inclement areas prone to flooding. But ergot is far from universally agreed as the cause, particularly given the marathon nature of the dances. Apparently, it would have provoked just a few passing spasms.

Research into climatic history is proving one of these theories right. We now know which points of the Little Ice Age were the harshest: namely the two hundred years from the late sixteenth century onwards. Though they ebbed and flowed, it had been 12,000 years since thirst and cold had had such a dramatic impact. And so mild had they been in the preceding centuries that people's initial reaction was to convince themselves that they had provoked God's wrath. Why? For having lived too well, too comfortably, too peacefully, and for being too intent on doing away with forests in times of plenty. If we go back to the years prior to the dance initiated by Frau Troffea, we clearly see the influence of the climate in the despair of that era; her suicidal dance was merely a taste of what was to come.

But the weather, as we know, never works in isolation. For the historian of medicine John Waller, the dancing mania was a pathological response to the fear and anguish of that time, something that would not have been possible without the religious pressure typical of a place where a "fear of god" was prevalent, fuelled by the belief in divine punishment. Geoffrey Parker also argued that another epidemic, that of the seventeenth-century revolts, was partly determined by the widespread tendency to focus on guilty sinners, whether oneself, the nobility or witches.

Troffea's century began with a ball of flame hurtling across the sky, which in those days would have been taken as an ill omen. In fact it was just a comet. From then on, however, the people of Strasbourg suffered cyclical droughts, frosts and epidemics that brought them to breaking point. Even so, by the spring of 1517, their hopes were somewhat renewed. It was going to rain, people said. But weeks passed and the rain didn't come, and by late April, when it seemed it must be about to arrive, when people were more certain than ever that it would, instead a frost came that nipped grapes and wheat alike in the bud. The population grew weaker, and was soon visited by the bubonic plague and the English sweating sickness. The latter was particularly harrowing, given that the physical effects included anxiety, dizziness, shivering, extreme tiredness, profuse sweating and panting, and led people to experience a terrible thirst that not even water could quench, before dying in blood-curdling fashion. It would be a matter of a few hours between its onset and death. Leprosy was also rife that summer.

For those who survived the intense cold, the failed crops and epidemics—and indeed survived the year they believed was going to mark the end of the world—the only remaining certainty was that nothing about the immediate future was certain. The streets were awash with prophets trumpeting impending apocalypse. Life was expected to be short and to end nastily. The population, which had multiplied and now had nothing to eat, was gripped with fear and a sense of hopelessness. Fiery balls continued to appear in the sky, while uprisings and rogations took place everywhere. The entire world seemed to be at war. If it wasn't a stifling drought people had to contend with, it was catastrophic flooding or gelid conditions. Before long, shortages drove people to take to the streets.

And then there was that weary, constant pain that had lodged itself somewhere in the body, which they tried to quell with certain vices. Melancholy, some called it, and while people tried to get to the bottom of that epidemic of depression and suicide, those affected turned to an array of avoidance techniques to make their lives bearable: sex or the cloistered life, gardening or addictions to substances—alcohol and opium particularly, though tobacco, coffee, tea and chocolate all had a moment too. Some of these were even occasionally used on prescription: given that melancholy was thought to be due to an accumulation of semen, doctors prescribed sex to stop body and mind from exploding. The church, meanwhile, urged moderation in eating, drinking and sexual relations, to assuage an angry God. As well as comets, the finger was pointed at sunspots (in this they were not wrong, only that the interpretation was back-to-front), as well as the theatre and dancing. Balls, theatre and skirts were variously banned in European cities.

Some sort of collective response was to be expected in cities like Strasbourg. In the years preceding the dance epidemics, the authorities and church leaders lived in fear of popular uprising every time "the heavens refused to open" and the price of bread went up in Alsace. They had feared it, in fact, since the sighting of the comet at the start of the century, because that harbinger of doom was interpreted in a very specific way: it foretold sedition. And they were partly right, because the rebellion had begun to be organized, although the man who had incited the poorest to rise up against the powerful profiteers suddenly abandoned them to their fate. He disappeared one day and was never heard of again. But the seed of his speech had already germinated among a population accustomed to hearing cries of imminent apocalypse in the streets. Nobody imagined that the response would come in the form of music only a few could hear. In other places, the expected reaction came.

In Spain, in the period after Frau Troffea's ball, simmering discontent followed three years of drought and was little improved when the king ordered the widespread requisitioning of wheat. A series of poor harvests had the usual consequences: grain shortages and increases in the price of cereals and bread. In 1521 the Green Banner Riots broke out in the Feria neighbourhood of Seville. On the instigation of a local carpenter, Antón Sánchez, Sephardic Jews, Romany and Moors, guild members all, as well as vegetable sellers and indentured serfs, went to the Corral de los Olmos, as they put it, "to demand an explanation": while riches continued to pour in through Puerto de Indias, Seville was suffering plague and an increasingly acute famine. The authorities tried to appease them by doling out wine, but it was bread they wanted. The green banner they carried, from which the revolt took its name, had been stolen from the Omnium Sanctorum and had apparently belonged to the Almohads, although its origin is contested. Promises were made, but the rebels sensed that they were bogus and went out in search of weapons. Four of the ringleaders were later captured and executed. This uprising, which was repeated elsewhere in Andalusia both that year and throughout the following century, apparently inspired the autonomous community's current flag.

Barely fifty years on, temperatures dropped by an average of 2°C, affecting ocean currents which in turn gave rise to extreme meteorological phenomena for more than 100 years. The consequences of this change in climate were global and included unprecedented famine in China and intolerable cold both in North America and across the Ottoman Empire. When the Little Ice Age entered its harshest phase, it wasn't just religious faith that people turned to: peasant rebellions broke out, while the foundations for the Scientific Revolution would be laid by the likes of Francis Bacon, René Descartes and Galileo Galilei, marking a change in the prevailing mentality. After seeing how severely Galileo suffered for his ideas, Descartes hid some of his writings. He himself fell victim to the Little Ice Age, as he is thought to have died of pneumonia when it was at its peak in the mid-seventeenth century.

A century after the Green Banner Riots, both Seville and New Mexico experienced their worst ever floods. At the same time the drought of 1628–31 was spreading famine and plague in places like

Spanish Lombardy, where a quarter of the population perished. The drought also decimated the population of Castile, although this was not the only cause. There, as in Andalusia, more grain was requisitioned by Madrid, where it was sold back to the rural population at twice the previous year's price. Rural depopulation was rife. In Barcelona, a ravenous mob seized bread from the ovens and devoured it before it had even risen. In Lisbon there was practically no grain, and in Biscay the women rose up against an exorbitant salt tax, while the menfolk looked on in silence. Catalonia, the Netherlands, Portugal and Aragón were on the verge of rebellion. The sun never set on that empire, and neither did the disaffection.

There were warnings, but they fell on deaf ears; neither Felipe IV nor the Count-Duke of Olivares wanted to hear them. Olivares not only ignored the signs, he also took advantage of the situation to launch a massive recruitment campaign in Castile, where there was barely enough grain to make bread. As a result, some Castilian villages were stripped of up to half of their inhabitants by a combination of climate catastrophe, military conscription and emigration.

When the drought came to an end in the Meseta, the respite was short-lived. A decade later it returned, only to be followed by Andalusia's heaviest recorded rains. From then on, Felipe IV surrounded himself with sorcerers, prophets and mediums, although he ignored any warnings that happened to be about the present or the imminent future. "It will all come down at once," they told him. In the ensuing years, Madrid was battered by torrential rain. In Andalusia, this time in Ardales, Malaga, there was an uprising in 1647. The rebels moved on to other cities shortly before yet more heavy rains destroyed the harvests. The price of bread tripled.

On 6th May 1652, a woman went weeping through the streets of Córdoba, a city that had just lost a third of its population to the plague. We know almost nothing about her, except that she was Galician, famished and had just lost a child. Rather than starting to dance, she simply walked through the San Lorenzo neighbourhood. As her name is unknown, we can let ourselves imagine that she was called Balloada, which is how Galicians refer to rain that comes on suddenly and intensely and then lasts for days. According to the chronicles of the time, Balloada walked the streets with her dead son in her arms. She had spent the morning berating the men for their inaction and cowardice. She was soon joined by other women, also unnamed, and a few men who went on to be immortalized in street names such as Calle Juan Tocino.

The people of Córdoba had been afflicted by years of relentlessly alternating drought and floods. The drought of 1651, a particularly bad one, had the usual domino effect: poor harvest, wheat shortage and rising bread prices. But they discovered that the climate was not their greatest enemy, that things were not as serious as they had been led to believe, and wheat not as scarce either; rather it was being hoarded by a handful of aristocrats and clergymen. And this was Balloada's reproach to them, as she held her son's body in her arms: their inaction and silence, when they knew full well what had caused his death. It was then, spurred on by the mother's grief, that the people decided to seek out those responsible. As soon as he heard about these unusual guests, who came not with a bottle of wine to share, but with clubs, sickles and scythes, the chief magistrate fled. Other profiteers, also sensing danger, followed suit. For several days, the rioters ransacked their houses and seized the grain. Diego Fernández de Córdoba promised them that the price of bread would be lowered and that there would be no reprisals. He managed to get himself appointed as the new chief magistrate.

The king issued a general pardon and sent more wheat from Madrid, at least comprehending that punishment of the rioters would only make things worse. But in another sense the lack of consequences backfired: it set a precedent. The people understood that they would get off scot-free if they took to the streets to seize grain. So the mob treatment of profiteers then spread to Seville.

What happened in both places became known as the bread riots, hunger riots or subsistence riots. But if there was no bread to eat it was because several years of drought had withered the crops, while those in power stockpiled grain, and taxes for the working classes kept going up. Like other rebellions of its kind, it was spontaneous and disorganized, for all that the rebels quickly rallied together. Food riots were a common occurrence in Europe between the fifteenth and nineteenth centuries, especially in grain-producing regions like Castile and France. Many of them were, at the same time, thirst riots. In Spain they reached their peak in 1766, in the wake of a four-year drought.

As if climate catastrophe, war and epidemic were not enough, the rural population of Europe saw taxes increase and new ones cooked up, this shortly after they lost their rights to the common land that had always been their bulwark in times of scarcity. The selfishness and indifference of some nobles created the perfect breeding ground for angry, hungry and often thirsty mobs to rise up.

It was a time when taxes continually rose, while poor people's children were enlisted for distant wars and soldiers were billeted in their homes. The clergy were exempt from such obligations, and anyone who amassed enough wealth to qualify for a noble title could also avoid them. Some regions were saddled with fixed taxes regardless of the number of inhabitants, so that they had to contribute more and more as their residents died or moved away. This gradual suffocation of places affected by depopulation saw them empty out almost entirely. The few remaining inhabitants became both more defenceless and more susceptible to sedition, especially when they saw so many of their fellow citizens taken away to serve in the wars declared by kings and waged at their expense all across the world, as well as by epidemics and famine. In this sense the uprisings were actually stirred up by those in power. Elsewhere too, the people knew that, while their own children starved, aristocrats and clergy were hoarding grain that they had been led to believe was in short supply.

It wasn't just in Europe that uprisings took place. Popular discontent had been on the increase since the end of the Middle Ages in almost every part of the world, whether that be France, Spain, the UK, Mesoamerica or China. Rulers everywhere were equally self-serving and cruel. While some kings and emperors exacerbated people's problems, however, there was at least one exception.

At the time of writing there are droughts in Gujarat, India. Not for the first time; in 1630 records state that it was impossible to walk through the streets without stepping on corpses, and that parents had begun to devour their own children out of sheer desperation. They were used to the monsoon on which they depended failing once every hundred years, but not four times, as had happened that century. The situation in Gujarat seemed insurmountable: when the drought ended in 1632, huge downpours laid waste to the land, depriving the population of grain and hastening the spread of malaria and dengue fever. Around a million people are thought to have died. Against the odds, the country recovered quickly. The emperor Shah Jahan took a very different approach to that of rulers elsewhere. He distributed much of his wealth among the poor of his city, giving out thousands of rupees every Monday for twenty weeks.

He opened soup kitchens and alms houses so that no one would go hungry. When the rains ended, he travelled to the worst affected towns and villages and distributed ten times more aid among the most destitute in rural areas to aid recovery. He also brought ploughs for the people, shortly before founding several cities, complete with markets, and putting efforts into an export drive. Although he spared nothing, for him it must have been a pittance because he still had money left over for an ostentatious renovation of his throne, to create the Shalimar Gardens and commission a colossal mausoleum for his deceased wife, this perhaps being the best-known building ever constructed in the name of love: the Taj Mahal.

Despite the waves of riots—like the 1775 *guerres de farines* in France—the monarchy was not always the target of people's ire. Their ministers became the scapegoats, while the praises of the kings themselves continued to be sung. "Long live the king, down with bad government," cried the Madrileños who took part in the Esquilache riots of 1766. The Marquis of Esquilache, the Sicilian-born Secretary of the Treasury, had ordered that the publicly acceptable outfit of long cape and wide-brimmed hat was to be replaced by a shorter cape and tricorn, which sparked protests amongst a people who felt that their identity was under threat. For a long time this unrest was put down to the capes and hats, and is usually therefore seen as distinct from the food riots. But there was more to it than that. The rioters soon began to demand cheaper bread, the price of which had doubled since 1760; successive droughts in the intervening years had left the fields scorched. Not only that, but other basic foodstuffs like oil and bacon had also shot up in price. Added to which, the government had taken steps towards economic liberalization, such as abolishing the grain tax, which threatened to make bread even more unaffordable.

Very much on people's minds in Madrid were the Cat Riots that had broken out fifty years before. In that case, a local woman, who could have been the grandmother of any one of these new rioters, had gone and lambasted the mayor: feeding her children had become an impossibility. "Get your husband castrated," the mayor shot back, "then you won't be spawning all these children." Those who heard this, a priest included, couldn't contain their ire, and a riot ensued. Carlos II was unwell in bed, rumoured to be dying. After various members of his retinue refused to go out on the palace balcony, the king got up from his bed and appeared before the people himself; once more the same old promises succeeded in calming the situation: prices would drop, the mayor would be replaced. The rioters then began to ask the king's forgiveness. And the king asked their forgiveness for having let them down.

In 1776, though, Carlos III felt frightened, took the uprising as an affront and demanded that the ringleaders be detained. Esquilache was dismissed and the Jesuits were also publicly blamed. The latter were an easy target, given that the Society of Jesus had just been expelled from Portugal and France. Finally, for a number of reasons, but supported by claims that they had instigated the unrest, the Jesuits were ordered to vacate the Iberian peninsula and the Spanish colonies in the Americas as well.

As well as being triggered by acceptable attire, the Esquilache riots in Madrid have long been seen as a catalyst for similar disturbances in the rest of Spain. The assumption that this wave of uprisings was due solely to the influence of a Madrid riot or changes in dress says a great deal about a country in which everything seems to revolve around an elite based in the capital, while the struggles of ordinary people and those on the margins are forever overlooked. In 1766 riots caused by a lack of water took place throughout the country, some after but some also before the Esquilache riots, mainly in the dry parts of Spain, although the wet areas were not entirely unaffected. People had been through four years of drought, bread was unaffordable and, in the face of all this, Esquilache decided to commandeer grain from rural places and carry it off to Madrid. Many of the uprisings took place around Eastertide. Drummers in the holy processions turned rioters. Like some surreal admixture of *Mission Impossible II*, Holy Week, the Las Fallas carnival and the San Fermín running of the bulls, that spring saw a confluence of drums, fire and knives in Spain. There were uprisings in Cuenca, Zaragoza, Barcelona, Seville, Cádiz, Lorca, Cartagena, Elche, Coruña, Oviedo, Santander, Biscay and Gipuzkoa. In the villages of La Mancha, pamphlets and threatening posters were distributed, particularly in Membrilla, El Toboso, Campo de Criptana and Granátula de Calatrava. The latter seems to have been led by angry poets: "Pedro Pablo, mend your ways/ cut the price of bread, or die at the stake." During the revolt in Cuenca, bread was handed out to the poor.

There continued to be uprisings, and neither did the weather relent. In 1780, crops again failed all across Europe, where economies had gone into freefall during the previous decade as a result of the credit crisis of 1772. On 8th June 1783, the Laki volcano in Iceland erupted. One hundred and thirty craters continued to spew lava for eight months, until 7th February 1784. The resultant noxious cloud spread across the world, and a topsy-turvy climate ensued. Summers turned cold, snow arrived late, droughts became more drawn out. Several years' harvests failed. Changes to river flow and long periods of drought led to multiple years of famine. Laki's eruption led to the deaths of six million people and countless head of livestock. Thirst was rampant, to the detriment of humanity. The successive droughts triggered by the volcano's eruption were decisive in the course of human history, as were certain thirsty women.

Stories told about Frau Troffea by those who knew her, although more level-headed than Paracelsus's misogynistic and prudish hypotheses, might still be nothing more than myth-making. Before she broke into her dance, for example, she was said to have thrown her son in the river to save herself. Myth or no, that same scene was repeated alongside uprisings elsewhere in Europe. In Thrace, similar stories were told centuries later, according to climate historian Emmanuel Le Roy Ladurie. Geoffrey Parker, for his part, points to infanticide as one of the most common measures used at that time to combat famine, both in China and across Europe; he also highlights the issue of overcrowded orphanages in various places, including Spain. The truth or otherwise of Frau Troffea's story aside, history is repeatedly marked with the scene of a mother weeping over her dead or starving offspring.

There was no single organizing factor to riots that arose during times of water scarcity, but they had more in common than just a historical period and the lower-class struggle to ensure their own subsistence. It was often women who set them off. Their involvement in various places in Europe had another underlying factor: they knew they wouldn't be subject to the same punishment as their husbands if found guilty of sedition. And indeed, their involvement was often forgotten about; save for the rare exception, they went unmentioned in articles and paintings, as well as when it came time to name city streets.

The weeping mother figured in both prayers and riots. In some villages in Valladolid, processions were accompanied by mothers in mourning—they wept in anticipation of their children's deaths, in the hope that the respective saint or the Virgin Mary would spare them. Elsewhere people rioted, lamenting their dead children and berating the menfolk for their apathy. A grieving mother taking to the streets was one of the most common portents of popular violence, coming even before satirical posters or the bells ringing out from the village church. Some of the women in question we have already mentioned—the anonymous Galician in Córdoba, the old woman who sparked the Cat Riots in Madrid—but there were others whose names did go down in history.

Josepa Vilaret spearheaded an anti-hunger, anti-establishment movement in Barcelona in 1789. On 1st March that year, a group of women entered the cathedral at the sound of the bells that were rung to alert the city to some impending danger; they were joined by a number of men, and before anyone knew it a crowd of more than 8,000 had gathered, instigating what came to be known as the *rebomboris del pa*—"bread disturbances". Josepa was among the first to be arrested. While ninety people were later punished with exile, she was one of just six to be executed; she was hanged in a public square alongside five men. Her fellow citizens refused to attend the spectacle.

That same year, in a Paris market, a boy started banging on a drum, and various women, fed up with bread being so hard to come by, set off towards the markets on the east side of the French capital and began ringing bells themselves. Other stallholders joined them, and alongside the revolutionaries, they descended on the Hôtel de Ville. Thousands upon thousands were there demanding bread. More and more drummers joined in, before one of them started shouting: "To Versailles!" The women led the way in that history-defining march. Drought was by no means the single cause of the French Revolution, but it was among the factors. Water had been scarce for a number of years, crop failure had become endemic, and the people of France were thirsty. On the day of the march on Versailles, it rained.

Things in Europe changed after that. But not entirely. In 1816, the year of no summer, groups of women in Britain spread the cry of "bread or blood", and in France there were bread riots once more. In Palencia, a woman named Dorotea Santos had a square named after her for her leading role in the bread riots there in 1856. In Madrid, the greengrocers on Calle Ruda and Cebada Market led an uprising—the Greengrocers Riot—in 1892. "May the rich/ have to eat their own tight fists!" went the chant. Famished, they none the less used vegetables as missiles. On a Monday in 1904, 200 Valladolid women—whose ranks quickly swelled to 2,000—descended on the offices of the local government to demand "bread and work". At the beginning of the twentieth century, those women were still calling for "the poor to get bread, and the rich to eat shit".

Owing in part to thirst, the idea of women as agitators had become widespread. The role of certain greengrocers in fomenting and sometimes spearheading the uprisings, because they were always out on the streets before people began to throng there, has left its mark on the dictionary definition of the term: apart from "person selling fruit and vegetables", in Spanish a "greengrocer" is also "a common, insolent person", and the term can be used as an insult. But nobody actually uses it like that. There is no doubt that people would have been proud to be known as agitators, greengrocers, and witches as well.

Aunt Casca lived in Trasmoz, Zaragoza, a place that was excommunicated in 1255 after constant struggles over irrigation water, had a curse placed on it in 1511, and in more recent times became a waterless "witch town". Nowadays the local inhabitants do not want the pope's curse revoked, and they have an annual "choose the witch" contest. But in the days of Aunt Casca, things were very different. People didn't even want to hear the word "witch" spoken on Fridays.

She dressed in the widow's weeds of one who has lost it all. She was a solitary woman, a wayfarer versed in plants. When she turned fifty-three, the town had been through its fair share of plagues and droughts. People started accusing her of causing the evil that was afflicting the land. Although the last auto-da-fé that saw a witch burnt at the stake was held in Spain in 1781 and the last woman accused of witchcraft in Europe was executed the following year, the idea continued to have currency, and Aunt Casca's fellow townsfolk decided to stage their own "trial" in the nineteenth century. They chased her, cornered her and threw her off a cliff.

After her death, local shepherds refused to set foot in the spot; they believed that Aunt Casca was unwelcome even in hell, and that her spirit wandered the land, her snake-like hair concealed under the skin of a wolf. One day, a young traveller arrived in the village and a shepherd advised him not to go along the Aunt Casca Camino. Intrigued by the fear he sensed, the young man wanted to know more. The shepherd finally agreed to tell him a story he claimed to have witnessed three years earlier, perhaps unaware that he was talking to someone who would rush to immortalize the story in writing. It was the writer Gustavo Adolfo Bécquer, who dedicated the sixth of his "Letters From My Cell" to Aunt Casca.

In it he quoted the Trasmoz shepherd who warned him against taking that path: "Coming to the cliff edge, she stopped for a moment, unsure which way to go. The voices of her seeming pursuers drew ever closer, and every now and again I saw her writhe, cower or scuttle a little way ahead, to avoid the blows raining down on her."

The woman, in the shepherd's memory, called for mercy from her accusers. According to him, the witch of Trasmoz said: "I am just a poor old woman who has never hurt anyone; I have neither children nor relatives to give me shelter. Pardon me, have mercy on me!"

But they accused her of a range of crimes none the less. They said she had taken away a mule's appetite. That she stole a baby from a cradle nightly. That she had hexed someone's sister. They accused her of putting the whole town under a spell.

One theory is that Aunt Casca was in fact Joaquina Bona Sánchez, a woman born on 10th March 1813, who married one Tomás Pérez and had four children. Perhaps this indeed was Aunt Casca and perhaps not; all that is known is that Bona Sánchez died in Trasmoz on 31st July 1860 of a sudden illness and was not given a Catholic burial. However, this does not match two accounts compiled at the time by Bécquer, one of which repeated Aunt Casca's claim of childlessness, while another, that of a young woman from a nearby village, said that she had a daughter.

Aunt Casca was not the only one accused by drought-stricken people. The worst centuries of the Little Ice Age saw a spike in accusations of witchcraft purportedly linked to failed crops. One of these stories took place very close by, in Épila. In 1631, a witch hunt was organized in the Aragónese village. Nine women were hunted down and imprisoned, and several of them executed. Carlos Garcés, who looked into the story, came to the conclusion that the arid conditions of the previous years were the principal cause of the Épila witch hunts. In his view, such persecution would be explained by a severe drought followed by floods, crop failures, famine and epidemics.

A minister in Russia recently put forward a madcap and apparently novel idea: weather reporters who got their forecasts wrong should be fined. But this isn't a new idea. The blame impulse has been part of humanity's story ever since somebody looked up at the sky and wondered if it was going to rain. Great power means great responsibility, and any intercessor is left open to accusation. From Lycurgus in ancient Sparta to Charles Hatfield in early twentieth-century California, those who have claimed to be able to summon the rain have paid the price when it either stays away or falls too copiously. Lycurgus and Hatfield suffered but in different ways: during times of drought and flood, the former was murdered (according to legend) and the latter was sued several times (in real life) but acquitted when the San Diego floods were deemed not to be his doing, but an act of God.

The Akkadian king Naram-Sin, grandson of Sargon the Great (founder of the Akkadian Empire), is the main character in a story that may or may not be true. He dreamed that Enlil, the principal god in the Sumerian pantheon, abandoned him and made the other gods of the city follow suit. When he woke up, he immediately went and put on mourning garb. Naram-Sin fell into a depression, unable to comprehend all the misfortune being visited on him. There was a seven-year dearth: grains and vines failed to grow, there were no fish to catch, and the clouds withheld their rain. Naram-Sin offered up endless prayers to the gods, but to no avail. He was too late: his unlucky fate had been written since birth. After seven years with his head in his hands, the king knew he would get no answer from the gods. Enraged, he raised an army and set off with clubs and axes for the temple of Enlil in Nippur. When he tore the drainage pipes from the building, "the water returned to the sky". He left the temple in ruins. Enlil then drove the Gutis down from the mountains and sent them to destroy Akkad. With them came hunger, ruin, despair and death. The city was destroyed. An empire barely a century old collapsed.

The story forms part of Naru literature, in which humanity clashed with the gods and thereby came to understand its own limits. It enjoyed great popularity in Mesopotamia although the stories couldn't be said to be particularly realistic. For all the historical details they contained, their value is primarily literary. The story of Naram-Sin's tribulations appears to have been invented some time later in the city of Ur and to have had didactic purposes, though we do now know that there was also a great drought in the Akkadian Empire at the time. The message contained in "The Curse of Akkad" is clear: there are consequences if one believes in a God of rain, but also if one seeks to punish the gods.

In a Roman myth, a people afflicted by thirst turned to an oracle in search of answers. And the oracle's response was unequivocal: the only way to end the drought was to put an end to the man responsible, who was none other than Lycurgus, the Thracian king of the Odrysians whose son Dryas had waged war on the followers of Dionysus. As punishment, Dionysus had sent drought to Thrace. To get the rains to return, an angry mob rose up and, on the oracle's orders, went to find Lycurgus. His subordinates handed him over for the horses to tear his body to pieces. Dionysus duly lifted the curse. And it rained once more.

Other rulers would have done well to know about these fictional tales. A recent study found that the majority of magnicides in ancient Rome had one thing in common: thirst. The very same thirst that drove Attila the Hun to invade Rome, according to a different study. In the middle of a drought, neither kings, rainmakers nor saints would be above retribution. Witches were also sometimes arraigned on charges every bit as flimsy as those that led to Lycurgus's assassination.

There were others, meanwhile, who comprehended that unless they punished themselves first, the day would come when their subjects would do the job for them. Around the same time that Roman emperors believed themselves the incarnations of Jupiter, the great Mayan kings in Mesoamerica considered themselves descended from the gods, and took on the role of rain gods on earth. The Maya believed that they all came from the water and "returned to the water"; this was their way of referring to death. Tikal, the largest Mayan city, was a long way from any rivers and depended entirely on rainfall. They built reservoirs there, enormous "mountains of water" that held millions of litres of water. They even made their own filtration system using stone and sand. The Tikal reservoir, which diverted rainwater into great cisterns, irrigation canals and smaller domestic tanks, had such a huge capacity that the inhabitants of Tikal could manage without rain for years. But the peasants in the surrounding countryside did not have the same luxury. In dry weather, their water holes kept them going for barely a few months.

To glorify the kings, stelae were carved with their accomplishments. These proliferated when the Mayan civilization was at its height, but ceased to be produced around 700 AD. A century or so later, Tikal was abandoned by most of its inhabitants; the rain had dried up, and the people ceased to believe in their king's ability to command it. On top of which, he had filled his coffers by increasing the tributes owed to him by the struggling peasantry. Centuries before the bread riots in Europe, popular uprisings erupted against a ruling class that promised non-existent rain. By claiming responsibility for the rain, they could expect gratitude but also castigation.

At the beginning of their mandate, but also when it came to annual festivals and rain-summoning rituals, the kings would stage a singular self-sacrificial act. They would appear before the crowd in only a loincloth and their finest feather head-dresses. They would then take a poisonous stingray spine from a bowl and start jabbing their penis with it repeatedly. The spilled blood was collected on pieces of parchment which were then burned in a censer while the king performed a frenzied dance. Evidence of the ritual remains in the form of stelae, friezes and paintings; in the ritual spines themselves, too, which held such importance that they were included among the leaders' grave goods. The stelae were frequently built, until the drought worsened and survival took precedence over the glories of the nobility. Discontent grew despite all attempts to quell it.

These self-mutilations in times of drought were aimed not at suicide or disfigurement, but to invoke the rain, with the added benefit of appeasing the people; the king could offer up his blood before they came for it themselves. The Dinka rulers in ancient Egypt, on the other hand, did go to such extremes: since they were forbidden to die of old age or illness, they knew that their time had come when they failed to bring rain to their people. Ineffectual rainmakers would have a pit dug in the ground and their belongings brought to them. There they would spend some time alone, before making their final wish: "Cover me with earth."

Impending starvation calls for desperate measures, but so does going thirsty. Although thirst has always been considered secondary as a driver for deaths, uprisings and migration, the body can go for anything between forty and sixty days without food, eleven without sleep, but only three without water. In normal circumstances, that is. In extremely hot, dry environments, twelve to fifteen hours are all that is supposedly needed before the body starts to shut down. Cases of people drinking their own urine to survive are widely documented. Alexander the Great's army were said to have licked beads of moisture off their swords. The Armenians expelled from the Ottoman Empire during the 1915 genocide set off with a pomegranate seed under their tongues; some of the survivors claimed they lived in the desert on a single seed a day. Hence the belief among their descendants that a pomegranate lasts 365 days, and their elevation of the fruit to a national symbol.

As we have seen, at the beginning of the Little Ice Age, people tended to see God behind any great misfortune. Self-blame was common, on the assumption that one had angered Him, but so was the idea that not all was lost because He who had the power to punish could also save. With the Scientific Revolution, there came an increase in rational explanations, along with the suspicion that there might be more at play than met the eye. With this, a former hopefulness began to fall away; perhaps not even God could save them from death by starvation. Some began to disbelieve the things they were told about the climate. In the final analysis, perhaps God wasn't so angry after all. Maybe the witches had cursed the crops. Or maybe the crops hadn't actually been that bad; maybe it was all down to a group of malignant, powerful hoarders. Maybe the real cause of their penuries had a name, and flesh and bones; maybe it was the ruling classes who with their apathy or sheer despotic nature had provoked so many catastrophes. Along with the already familiar enemies, like the stars, the fiery balls in the sky, eclipses and people's sinning natures, now there was a whole new set of scapegoats, who were in fact the scapegoats of yore by another name: the nobility, witches and rainmakers.

The idea that God had deserted humanity began to take hold, and the ruling classes were considered to have done the same. With the loss of hope came a loss of faith, and in turn a reduced fear of the consequences of sedition. If rogations did not bring about a miracle, guilty parties would be sought. Much of the population had lost faith in their rulers, whether they were in Tikal, Strasbourg, Córdoba or Paris. Science and reason did not do away with the saints but, like the ruling classes, they too came in for punishment. Yet science, and specifically meteorology, which in the seventeenth century was taking its first steps, were undoubtedly a great help to the intercessor saints and those who had not lost their faith. Anyone with access to a weather report—and it is from this era that they date—can have a pretty good idea of the best day for organizing a rogation.

When a rogation failed to have the desired effect, the parishioners took measures in accordance with the material from which the saints were made. With no blood to spill, and as a way of both making sense of things and punishing the saints who hadn't given them what they wanted, they began a tradition of dunking their heads in the water, or filling their mouths with salt or salt cod, or flinging the statues into the water. In Sicily, in such cases the saints would be buried, while in Japan they stuffed them with rotten rice. Few people are disposed to acknowledge acts of collective sacrilege despite the evidence—among these, the ritual of immersion seems to be most common.

But there is one village at least in Aragón where such practices are out in the open. The one I'm living in. People born and raised in Castelserás have been given the sobriquet of "Jews". And the explanation lies in one of the punishments that people don't tend to own up to, except here. Although there are various versions of the tale, if you ask around in the village the most common is the following (and although here I am writing about thirst while rain is falling outside, the underpinning for what follows is that it virtually never rains here, and that when clouds do come along they tend not to be the kind with rain inside them): there was a moment in the past when people were feeling so in need of rain that they brought a Christ figure out for a procession. In his memoirs, Luis Buñuel, a native of the neighbouring village, talks about its having been the Virgin Mary rather than a Christ. Suddenly the sky grew dark and a sense of hope suffused the gathering. But just as suddenly, the clouds played a familiar trick, veering off in a different direction. It seems that the people took this as mockery on the part of the intercessor, and went and threw the statue in the river to teach it a lesson. If this truly happened, it took place on the bridge I can see from where I am writing at this very moment. The figure was washed down the Guadalope River, until some men from Alcañiz came across it. The only possible authors of such a sacrilege could, in their view, be Jews. And from that day forward, that was what they called them. According to another version, there was no Christ figure in the water because in fact it was a sack of straw that had been tossed into the river, as a way to bait the Alcañiz men. What started out as an antisemitic slur was then reclaimed by the people of Castelserás, who now take pride in it.

Whether this is a legend or an event that truly occurred, there had to be sufficient trauma for it to be consigned to myth: a bridge had to have been built barely two or three centuries earlier, which the water then rose to cover in a place with a markedly damp climate when the dinosaurs roamed, but where now there was little but dust. People had even once gone in procession to the Fórnoles chapel and found the members of seven other villages there, all on their own rogations and all having arrived with no prior coordination. Everyone thought it was a miracle and decided to repeat the gathering, until it became an annual festival that has lasted for centuries.

10

Feet on the ground, eyes on the sky

> To understand life in its totality, one has to look both up and down, at the thick boughs and slender branches, the crown of the tree and the tindery fallen twigs; sweet-smelling and rotted-down, present and past.
>
> JUAN LUIS ARSUAGA

"If you're so wise, why aren't you rich, why don't you prosper instead of being a total failure?" One of the Seven Sages of Greece, who was the first-ever philosopher and a forerunner to western scientists, often had to contend with this question. In a world that was awash with myths, he was the first person—the first recorded person—who sought to understand reality on the basis of rational thinking rather than fiction. He searched for an origin to all things and found it in water, which with its divine essence seemed to him the habitat of the gods. He considered it a living being for the way it moved, which also meant it had a soul. He also believed that the Earth was a kind of island floating in the water and that this explained earthquakes. Herodotus recounted, although he himself seemed sceptical, that the man was said to have split a river in two.

He learned to read the sky, predicted the timing of an eclipse and took measurements of the moon, and of the pyramids as well, on the basis of his own shadow. People often went to him for advice, but mocked him when he decided to buy up all the olive presses in the city in the middle of a drought, at a moment when the crops already seemed certain to fail. Once, an old woman who had received him at her home took him out onto her terrace to talk about the stars. The philosopher, eyes fixed on the heavens, tripped and fell into a hole. The old woman, who couldn't contain her laughter, finally managed to say: "Oh, Thales! Do you really expect to know the secrets of heaven when you don't even notice what's right beneath your feet?"

But he knew that the firmament was full of encrypted messages. That's why his knowledge of astronomy allowed him to predict rain when nobody was expecting it. And he hadn't actually forgotten about things at ground level, because he had been studying the olive trees and had calculated how much water and sun they needed; this was what led him to buy up all the olive presses in Miletus. When that year's olive trees yielded a bumper crop, his neighbours had no choice but to go to Thales to process it all and pay the high price he demanded. Thales was now a wealthy man, but he soon sold the presses and went back to his old life. In reality, he wasn't interested in getting rich, and even less so at other people's expense; he just wanted to show his neighbours that being a philosopher or studying was not some kind of failure. He'd had enough of hearing them say that he was wasting his life.

That man who looked to the sky for answers died of sunstroke during an Olympics. Although there are those who think he wrote about astrology, the solstice and the equinox, others believe that none of those writings were really his. If Thales of Miletus left anything in writing, no one has found it. Fortunately, Aristotle and Seneca ensured his legacy.

Although "meteorology" means the "science of the atmosphere and meteors", the ancient Greeks called everything that was up in the sky "meteors". It was probably Aristotle who coined the term in around 350 BC, when, after *On the Heavens*, he wrote his *Meteorology*, both of which were included in a four-volume compendium known as *The Meteorological Works*. Aristotle knew that rain was not the divine semen of Anu nor the breastmilk of Antu, his consort, as the Sumerians and Akkadians believed. He was probably the first to talk about the water cycle. Although he didn't get everything right, he was the first to leave written descriptions of the rain and its origins. His fellow Athenians went on imploring the gods, and were still doing so 200 years later. As recorded by Marcus Aurelius, they would repeatedly say: "Send us rain, send us rain, beloved Zeus, upon our fields and plains." "Either we must not pray, or we must pray like this, simply and spontaneously," added the Stoic emperor. But rather than pray, Aristotle elected to look to the sky. He strived to understand, and arrived at conclusions that were not always correct, but not always wrong either. He believed that a perpetual current flowed from the centre of the Earth outwards and back again; that water evaporated, rose to the sky and fell in the form of rain. He also intuited that the sun's halo was an optical effect, and that clouds were actually composed of water. That is why, in his ordering of meteorological phenomena, he classed certain ones as being "apparent", while rain he placed in the "real" and "immediate" category. Aristotle already thought, like the Mixtec, that clouds did not actually form in the sky, but on the ground. He wrote that rain "is formed from a large [amount of] vapour that cools", and that this came about according to the space and time "from which and in which it accumulates".

Though not the first to look for answers in the clouds—Anaxagoras had previously claimed that rain fell when the moisture in the clouds, flying too high, froze and was then forced by its own weight to descend—Aristotle was one of the first to write about the origin of rain: "Since moisture always rises upwards due to the effects of heat, and falls back to earth due to cooling, the names of these phenomena and some of their variations are aptly applied: indeed, when [the moisture] moves in small particles we call it a light mist, while when [it does so] in larger particles we call it rain."

Just as Thales of Miletus had done, Tirtamos, one of Aristotle's followers, focused not on myth but on observing signs in nature to predict the weather. It was Aristotle who decided to rename his favourite disciple Theophrastus, "for the divinity of his eloquence". In *De Signis* Theophrastus compiled some 200 omens, eighty of which were rain-related. This included one that is perhaps most widely cited in various cultures preoccupied with rain: the sun halo, which we will come back to.

When the Arabs brought their pre-scientific knowledge of the weather to Europe in the Middle Ages, they also brought Aristotle's *Meteorologica*. The Italian translator Gerard of Cremona was tasked with translating the work into Latin at the Toledo School of Translators, and from there it spread to the rest of the world. It became a compulsory text in universities and was unquestionably the most widely read treatise on physics until well into the seventeenth century.

Almost nothing more was written about meteorology until, in the thirteenth century, Albertus Magnus discussed the roundness of raindrops and, well into the seventeenth century, Descartes included an appendix in his *Discourse on the Method* entitled *The Meteors*, in which he talked about the whiteness of certain clouds being the result of accumulated droplets reflecting the sun's rays. He claimed that clouds contained not only water droplets but also ice crystals which collided, melted and fell in the form of rain. Until Galileo's time, Aristotle's ideas were the most widely accepted in western civilization, while others—such as the notion that scientific experiments were a kind of witchcraft—also remained current. Aristotle believed that the Earth was stationary and that the sun attracted moisture. But without going back to him, to Thales of Miletus, to Theophrastus, to the myths that preceded them and to the sorcerers, magicians, priests and diviners of the ancient civilizations that in turn allowed for the early glimmers of meteorology, there would be no way to understand the changes that took place during the darkest moments of the Little Ice Age.

From the seventeenth century onwards, the empiricism of those who laid the foundations of pre-scientific meteorology was no longer enough. People realized that appealing to the sky, the gods, saints and sorcerers for water didn't work. They also realized that meteorites, floods and droughts were not divine punishments, mistakes on the part of sorcerers or rainmakers, or the work of witches. However, things didn't change overnight. Invocations and scientific advances were not mutually exclusive. At the time of writing (2023), after months with no significant rainfall in the dry parts of Spain, reservoirs are completely empty following the "April showers", flamingos are on the verge of giving up on Doñana, a traditional breeding ground, and relocating to Albufeira, and people everywhere are praying.

But let us go back to a time when a select group of people were obsessed with the skies, with the air, with water and the immediate future. Some of the moon's craters were named after them.

Predicting storms, making sense of sunsets and sunrises, identifying snowflakes, measuring raindrops and the intensity of the blue of the sky, and creating a system for clouds were some of their main achievements. These passionate, dedicated individuals enabled humankind to know about bad weather in advance, to tell different kinds of clouds apart and measure rainfall. Meteorology was still included within physics, but it was soon to fly the nest and become a discipline in its own right.

We don't know for certain when it was that humans first started measuring rainfall, but there is evidence of the activity in Jericho some 8,000 years ago. In any case, just predicting the rain didn't cut it; it had to also be measured and, if possible, stored. Although measurements continued to be taken in different places such as ancient Greece with archaic instruments, there is no evidence of rain gauges until the fifteenth century in Korea, in the time of Sejong the Great. It is precisely to him and his son Munjong that the invention of the *cheugugi* is attributed, the first known rain gauge in history, which was then adopted in other parts of the world.

As we've seen, Descartes was one of those people focused on the sky from the mid-seventeenth to the nineteenth century, posing questions about the nature of rain and offering an explanation for the whiteness of clouds. In 1643 Evangelista Torricelli invented the barometer. Just a decade later, Ferdinand II of Tuscany too got the meteorology bug and ordered the construction of Europe's first network of weather observatories; he also founded the Palatine Meteorological Society in Mannheim. Although the Italian physicist Benedetto Castelli, a disciple of Galileo, is credited with making one of the first rain gauges, Christopher Wren improved on his efforts, just a few decades after Robert Hooke invented the pluviograph to measure the intensity of rainfall.

For his part, John Tyndall believed that we live inside the sky and not beneath it. Two centuries after Descartes mused on the whiteness of clouds, Tyndall asked why it might be that the sky is blue. In the intervening years, Horace-Bénédict de Saussure, one of the founding fathers of mountaineering, invented the cyanometer, which measured the colour intensity of blue sky. With the help of a hair, Saussure came up with the hygrometer; if moisture made it ripple, there was nothing like a taut strand to measure the degree of ambient humidity. First with human hair and later with horsehair, his invention remained in use for centuries.

Although other tools were devised for measuring rainfall, in the nineteenth century the meteorologist Gustav J.G. Hellmann invented the rain gauge that was eventually patented by the World Meteorological Organization. Unlike its predecessors, Hellmann's device retained all the water it collected, losing none to evaporation or the impact of falling droplets.

So the barometer, the rain gauge, the pluviograph and the hygrometer were all invented during the harshest moments of the Little Ice Age. But more still needed to be known about rain. Clouds, for one, whose current names we owe to an Englishman who spent the summer of one year without a summer gazing at them. Today we know that rain comes mainly from two types of cloud: stratus and cumulonimbus. Nimbus means "rain cloud" in Latin, while cumulus means "heap" or "mass". The first are grey, dark. The second are shaped like whipped cream and rise to a kind of anvil at the top. The reason that we now call them this is that someone devoted enough time to them to understand that they could be classified and, therefore, named.

The summer of 1783, when Luke Howard was a child, was one of upheaval. Sudden changes in the weather set alarm bells ringing around the globe. An ash cloud meant the sky was constantly changing colour. Animals died and crops spoiled for no apparent reason. In England they called it the sand summer. While devastating volcanic eruptions took place in Iceland, a volcano in Japan awoke as well, and in Italy the ground trembled. Perhaps it was then that Howard first began looking to the sky, because that was what his elders would have been doing.

Like so many of his contemporaries, Howard studied the sky to see if it looked like rain. By doing so repeatedly, he began to discover different combinations of water droplets, which looked to him like either wool, cotton or tattered fabrics. At the time, people were becoming intensely interested in nature and how it impinged on humanity, leading to expeditions to find new species and also to the development of taxonomies and order. The ever-changing clouds had so far been an exception. But Howard noticed that the clouds, although they changed day by day, hour by hour, and minute by minute, were often the same, or at least similar enough to one another to form a pattern, whether those be unicorns or heart shapes. Essential forms and textures kept reappearing. As a result of his observations, he established three types of clouds (cirrus, cumulus, stratus) and several intermediate categories that are used in cloud taxonomy to this day.

To follow the movement of the clouds and make sure he captured all their possible variations, Howard often travelled from London to the Lake District, where he would sketch them as they changed in different environments.

At almost exactly the same time, Jean-Baptiste Lamarck created his own classification of clouds, but his work enjoyed little renown because he was working in French. Howard, on the other hand, decided to establish a nomenclature in Latin. He presented his essay on cloud formations, also in Latin, to a group of researchers and thinkers in London in 1802. For Howard, clouds altered in accordance with atmospheric variations. According to him, they were unquestionable "visible indicators" of these atmospheric changes, just as a person's expression can be suggestive of their state. "In order to enable the meteorologist to apply the key of analysis to the experience of others," he wrote, "as well as to record his own with brevity and precision, it may perhaps be allowable to introduce a methodical nomenclature, applicable to the various forms of suspended water, or, in other words, to the modification of cloud." A year after presenting it, he saw the essay published in three parts.

His classification was to become known beyond the confines of science, and had special relevance in the arts. In the paintings of Constable and Turner there was a marked change in the depiction of clouds after Howard introduced his cirri, cumuli and strati to the world, which came accompanied by his own sketches and watercolours. His influence is also there in the poetry of Goethe and Shelley. He was in regular correspondence with the former, and they shared confidences; Howard, a pharmacist by trade, confessed to the German poet and naturalist that his true vocation was meteorology. Goethe took it upon himself to circulate Howard's essay, and some of his poems were inspired by the Englishman, whom he called "the man who distinguished cloud from cloud". Almost overnight, the meteorological enthusiast became a leading light, widely referred to as "the godfather of clouds".

Richard Hamblyn, in *The Invention of Clouds*, shows how "Goethe's admiration and his sense of indebtedness to Howard's meteorological theories... led to one of the most extraordinary personal homages ever paid by one scientific worker to another". The homage consisted of adapting Howard's now famous essay into a sequence of poems, one for each cloud type, the title of which made its import clear: *In Honour of Howard*. And he didn't stop there, but even managed to convince the Englishman to write an autobiographical text describing how he had come to give the clouds their names. Goethe received the autobiography and by the next day was already sure that it was the most delightful thing that had happened to him in a long time. Between 1820 and 1825, still enthralled, he wrote and drew in a diary the details of his own observations when he looked at the sky. It was the seed of his book *Cloud Poems*.

But winning admirers always means gaining detractors too, and so it proved for Howard. If until that time there had been a lack of interest in naming the clouds, it was because they had been considered unclassifiable, wild, ethereal, ever changing. This was Caspar David Friedrich's view when he expressed his incomprehension at someone seeking to confine the clouds inside the cage of order. Naming them on the basis of patterns meant destroying their "expressive potential". It was completely implausible and unnatural. Others criticized the fact that Howard named the clouds in Latin and not English. On this count, Goethe came to his defence, asking that the Latin go untranslated, otherwise "the first intention of their inventor and founder would be destroyed".

At the time Howard's star was rising, Robert FitzRoy was born in Suffolk, England. He went on to become commander of HMS *Beagle*, and was looking for a companion in his adventures for fear of the destructive effect of loneliness, having heard of numerous suicides among solitary sailors. He longed for a dinner partner, someone with whom to converse about nature, a friend. He ruled out several interested parties and finally selected a young man who applied when FitzRoy had all but given up hope: Charles Darwin. Together they set off to discover the world's oceans, and one summer's day as they were crossing the Atlantic they met a *pampero* wind. It was not FitzRoy's first experience of this strong, sand-laden wind, driving swirls of butterflies and dragonflies before it. He already had some knowledge of it and the storm it presaged. And bad memories, too. Just three years earlier, he had seen two of his crew swallowed up by the sea. He had been plagued by a sense of responsibility ever since, blaming himself for his failure to interpret the barometer, which had sent him warning signals in a series of abrupt drops. He had resolved that no one else should have to witness such a tragedy and became obsessed with finding a way to predict the weather. On his travels, he would take note of meteorological data to make navigation easier for others. His endeavours were interrupted when he was appointed governor of New Zealand. But he was dismissed from the post after just two years for failing to punish a Maori group after a conflict with settlers, and for having gone so far as to defend Maori land rights.

From then on he was able to devote himself to meteorology. He devised a storm warning system, drew up charts of prevailing winds, created a network of observatories that was the forerunner for the UK's Met Office, and when the Board of Trade's Meteorological Department was founded he was appointed its president. FitzRoy was not only a pioneer of meteorological observations. He published *The Weather Book: A Manual of Practical Meteorology* and also the first weather forecast in a newspaper; indeed he is said to have coined the term "forecast". With his forecasts in *The Times* from 1861 onwards, meteorology was popularized, coinciding with the end of the Little Ice Age. Weather reports—until then the preserve of an intellectual elite—went on to gain mass appeal through newspapers, and it was FitzRoy's doing.

Following the publication of *The Origin of Species* and the ensuing controversy, FitzRoy was overcome with guilt: a man of deep religious faith, he felt he had somehow contributed to the work. He fell into a deep depression and finally, in 1865, ended up doing the very thing he had tried to avoid by having Darwin join him on board HMS *Beagle*. He took his own life.

Although the Little Ice Age had come to an end, there was still a pressing human need to predict the rain. Now both the amount and intensity of rainfall were a concern. But to truly manage it, a fuller understanding was required, and that meant looking at the past, the present and the future, at the sky and the earth, as well as utilizing instruments that, although centuries-old, were constantly being perfected. Thousands of years passed between the water tank that was used to measure rainfall in Jericho and the construction of the network of meteorological radars, satellites and mathematical models. It was in the nineteenth century that numbers came into play. Since then, and especially since the beginning of the twentieth century—assisted by radars and satellites—mathematical models have been relied upon to predict the weather.

Even with all these advances, meteorology did not become a science in its own right until well into the twentieth century. The Spanish Meteorological Service, for example, did not exist until 1913, and before that "weathermen" were not scientists but fortune tellers, naturalists and Jesuits—in that order. The last of these, in fact, had built the first Spanish observatories in Manila and Havana some time before. There were very few women meteorologists until the middle of the last century, and their increase coincided with the use of weather satellites. Felisa Martín was the first woman to join the State Meteorological Agency in Spain and also the first woman in the country to obtain a PhD in physics. But there was one unknown woman whose work we have indirectly been discussing. When meteorology began to apply mathematical models, the meteorologist

Edward Norton Lorenz tried to generate computerized predictions in 1963. He concluded that an initial variation could lead to major differences in the results. Lorenz, father of Chaos Theory, popularized his own research with a beautiful metaphor: a butterfly flapping its wings could cause a hurricane on the other side of the world. At the end of his book, he did something that was far from customary among his contemporaries. He wrote: "Special thanks to Miss Ellen Fetter for taking care of the many numerical calculations."

But who was this mysterious girl? Ellen Fetter was a mathematician and one of the first ever computer programmers. She was only twenty-three when she started working as Lorenz's assistant at MIT. Fresh from a mathematics major, she was hired to replace Margaret Hamilton, who was later to take part in the Apollo mission that put a man on the moon. In fact, Hamilton herself recommended Fetter, who then took charge of the computer work, which consisted of plotting the movement of a particle undergoing rapid convection in an imaginary beaker. The butterfly idea came precisely from her work as the meteorologist's assistant, as the strange attractor diagram they produced together struck them as both familiar and evocative.

Human beings have always looked to the sky, often with yearning, to see when the rain might come. The overwhelming ambition has been to manage the rainfall, since the advent of agriculture in particular. "If the sky darkens three days in a row, expect clouds and rain in the following months," said a Babylonian omen. In Babylon they used haruspicy, a form of divination based on the inspection of the entrails of sacrificed animals, as well as auguries related to the flight of birds, and lecanomancy, which meant predicting the future according to patterns of oil within water. All of this fell to the *barû* priest, a futurologist who might rely on the observation of anything from animals to the plant kingdom or the stars.

Although meteorologists use mathematical models, radar and satellites to predict the weather nowadays—with varying but almost never total accuracy—there are still those from Theophrastus's school who, based on the clouds, sun halos, animals and plants, make the *cabañuela* weather charts by extrapolating from the first twelve days of the year, as well as daily paintings and almanacs to predict rainfall. Farmers, like shepherds, have always been inherently connected to nature and for centuries have made weather forecasts for the following year based on their observations of what happens in the sky, in the air and on the ground. They note details about the sunrise, the wind speed, the shape of the clouds and the moisture in the air during the first twenty-four days of August, the first twelve days of January or at midnight on 31st December, depending on where they are observing from. The origin of the *cabañuelas* has been traced back to Hebrew culture; they appear to be linked to the years their ancestors spent in the desert. During the festival of Sukkot (from the Hebrew for "huts", "tabernacles" or "booths"—hence the Spanish "cabaña" prefix), Jews build a hut and live inside it, for seven days in Israel and eight in the diaspora, to connect with the experience of their desert forebears. Its final day, known as the "day of the rain's judgement", is often wet, as the festival comes immediately before the rainy season in Israel, and people spend it praying for the rain to last through the rest of the year. In reality, Jews do not need to wait until then to pray for rain, given their desert roots. In their most important prayer, the Shema Yisrael, rain is described as a gift from God, reserved for those who fulfil the precepts and includes drought as a threat for those who do not. There was a time when the last day of Sukkot was used to pour water as an invocation, with the help of rainmakers like Honi, "the circle drawer" according to the Talmud.

But back to *cabañuelas*. For all their implausibility and lack of scientific basis, and the fact that they are increasingly difficult to get right, there are still farmers who keep track of data over the summer in order to prepare the following year's *cabañuelas*. Both my father and Pascual, the last farmer in the Elche countryside who still makes *cabañuelas*, showed me the process, but I have never had the patience to make one of my own. I belong to a generation with access to the latest weather forecast on our mobile phones, updated every two hours, not only with data from meteorologists but also with the help of algorithms and information sent by users in real time.

But methods very similar to *cabañuelas* exist with different names both elsewhere on the Iberian peninsula and in other parts of the world. Just as in northern Spain they have their *tempora* and *cuarta* weather charts, which correspond to the seasons and are based on the Catholic liturgical calendar (even though they predate Christianity), the Mixtecs still have their "painting of the days", while in the past there were the Mayan *Xook k'íin*, which literally means "weather readings". In both places, as well as this type of forecasting, the use of almanacs is recorded (the Galván and the Zaragoza calendars), despite the fact that even current-day technologies are unable to predict so far in advance.

Mariano Castillo y Ocsiero was known in the mid-nineteenth century as the Spanish Copernicus. He was born in Villamayor de Gállego (Zaragoza), but went to Cádiz to train in astrology and meteorology. There he worked at the San Fernando observatory, and later moved to Madrid before returning to Zaragoza. In all those places he devoted himself to observing the sky. He took daily notes on what was happening around him: air pressure, temperature and the appearance of the moon and sun. By combining his own observations with newspaper weather reports, he detected predictive patterns and began to publish *The Firmament*, its subtitle "the only legitimate almanac for Zaragoza".

His calendar is still popular today. Although his face appears on the cover of the almanacs that are sold in bookshops and stationers, he did not choose its name because of his own origins, but because his region had produced almanacs historically. For a long time it was believed to be a tribute to an Aragónese astronomer, Victoriano Zaragozano, who was already producing this type of almanac in the sixteenth century. Joaquín Yagüe, Mariano's predecessor with his almanac *The Heavens*, was also from Zaragoza. When *The Firmament* was first published in 1887, which was before the founding of the State Meteorological Agency (AEMET), over a million copies were printed. Its design has not changed over the years, so buying it is like preserving a family tradition. It is like acquiring a time capsule that, as well as taking you back to the past, allows you a glimpse of the future.

Such was the precision of Castillo y Ocsiero's forecasts that his neighbours decided to play a joke on him one day that is still talked about anecdotally. They placed cigarette paper under the platform he went and sat on every day to contemplate the stars, and Castillo y Ocsiero said as he sat down: "Either the earth has grown bigger or the sky has got lower". They say that on one occasion he predicted a heavy hailstorm and also his own death. That day, he asked his wife to take him home to die in peace. After his last breath, the hail arrived.

Nowadays a pair of fairly discreet brothers produce the almanac, apparently based on data provided by the Madrid Astronomical

Observatory. Although Castillo y Ocsiero has been dead for over 150 years, hundreds of thousands of copies of his almanac are printed and distributed throughout Spain every year.

Heirs to Theophrastus still live on in rural areas, relying on direct observation of the sky, animals and plants, and also on proverbs. My grandparents' generation repeats rhymes that refer to a specific date or saint, as well as to natural phenomena that allow us to know what is in the offing. Many of the observations passed down through oral tradition share a starting point that science would be better not to ignore or ridicule, because popular wisdom is not the enemy of science. Sometimes it can even be its ally; an empirical conclusion can just as well be a scientist's starting point.

The sun halo may be the most globally known predictor of rain. The idea that it precedes rain has been present throughout history. One of the smaller stelae from the library of the Assyrian king Ashburnipal held in the British Museum states: "When there is a small halo around the sun, rain is coming." There is no great difference between Jordanian proverbs and the kind of refrains my grandparents would roll out, such as "Ring around the sun, shepherds best run." Paremiology, the collection and study of proverbs, has found up to twenty-three ways of explaining this omen in the Iberian peninsula alone. When the sun has a halo, it is usually red with a purple rim, while the moon's is usually white. The halo, as Aristotle already realized, is an optical phenomenon. It occurs when we look up at the sun through cirrostratus clouds that contain particles of suspended ice; these types of clouds bring rain. And this isn't the only instance in which weather lore has got in ahead of the science.

"Mackerel scales and mares' tails make lofty ships carry low sails" is an example of weather lore that has been backed up by science. Here the altocumulus cloud is being described, which though not itself a rain cloud, often appears on the eve of a storm, knowledge that has been around since ancient Roman times. Although these popular predictions are often labelled as superstitious, unscientific and inaccurate, oral tradition had already prepared our grandparents for such criticism: "Fibber, fortune teller, birds of a feather."

In the past, it was not only by looking at the sky that one could read the weather. To know if rain was approaching, the Mixtec people look to other signs, such as the dampness of the earth, birdsong and animal behaviour. They take guidance from the croaking of frogs, the presence of "water grasshopper", the downward flight of fireflies and increases in the number of army ants. The last of these is not mere superstition either, as army ants have been shown to change their behaviour when rain is approaching, being able to sense ambient wetness before humans can. These insects hold perhaps the highest importance for the Mixtec, and are considered intermediaries between heaven and the underworld due to their ability to both make nests underground and to open their wings and fly. It wasn't just rain that these ants announced: when their bodies turned light brown in colour, a shade similar to coffee, the Mixtec took it as a warning of a coming drought.

Like ants, bees are far more sensitive to water in the atmosphere than we are, so rather than predicting rain we might say that they sense its approach. In Terrinches, people still remember an old man who used to predict rain based on the behaviour of the bees. He knew it would rain when he saw them in a state of agitation. Bees become hyperactive days before it rains. This is because, thanks to their hydro-receptors, they can perceive minimal amounts of moisture in the air, so they start to work frantically in order to gather nutrients and water for the hive. Shortly before rain starts, they will enter the hive and not come back out until it stops.

The Austrian biologist Karl von Frisch succeeded in interpreting the communicative dance of bees. His discoveries won him the Nobel Prize in Medicine in 1973. Folk wisdom: 1–nil. And to thirsty bees as well, which depart for wetter places at the end of summer, using their dance to inform the water carriers where the water is to be found. According to some studies, this could be due to the high level of sugars in their stomachs. Von Frisch himself created experiments involving water and sugar before a storm, although another study by the Agricultural University of Jiangxi in China showed that these insects change their habits before it rains: they work harder and return to the hive later.

In summer, bees need more water to bring down the temperature inside the hive. The forager bees go off in search of water, on a journey that may cost them their lives, as their wings have a tendency to stick to puddles. To avoid this, they do something akin to surfing. When they set off on their quest, they use their proboscis to suck up drops of dew. On their return, they regurgitate the water into the nurse bees (these are newborn bees whose job it is to look after the queen until they are twenty days old, when they become foragers) and place another drop in the upper part of the honeycomb cells. The drop evaporates and keeps the young bee from dehydrating and dying of thirst in summer. Bees are always thirsty. Forager bees can make up to 100 trips a day in search of water and are capable of carrying up to eight per cent of their body weight in water, more or less eight drops per trip. When they need to collect water, the forager bees become especially agitated. This state of hyperactivity will last for several days. When they go back inside the hive, it is one of the surest signs that rain is on the way.

✦

The Hittites believed that a bee could save us from going thirsty. According to one of their myths, Telepinu was not the most pleasant person in Hattusa, the capital of the Hittite Empire. By all accounts he was insufferable and he hated everyone else in turn. This god-king, son of storm-god Tarhunna, then disappeared without trace one day. Tempting as it might be to imagine that he made a smokescreen because he was so sick of everyone, this is neither the beginning nor the end of his story, which, though recorded on several tablets, has yet to be found in its entirety. What we know is that Telepinu disappeared very abruptly one day, dropping several balls at once; the springs all dried up, and the trees and pastures too; all the crops withered and people stopped bearing children. Humans and gods alike felt the pangs of hunger. "The great sun god organized a feast and invited the thousand gods," the story on one tablet goes. "They ate, but their hunger was not satisfied; they drank, but their thirst was not quenched." Telepinu withdrew from the world and allowed it to dry up, while his people starved.

It is likely that he took with him the bull habitually ridden across the sky by his lightning-bolt-wielding father; the same bull we have already discussed, bringer of both drought and rain. On the same tablet, Telepinu is accused of ruining the harvest with his departure, and of deforming the ox by taking it away. Not even the thousand gods there present could get Telepinu to come back. Sending an eagle to find him did no good. The mother goddess Hannahanna decided to send her bee instead, despite Tarhunna's mockery. She asked the bee to find Telepinu, to sting his hands and feet, to purify him with its wax and then to bring him home again. Thus, a bee was entrusted with putting an end to a devastating drought, and only the mother goddess believed in it. And that belief proved well founded. The bee stung Telepinu's hands and feet as instructed, but Telepinu woke up just as angry as he had been before. Angrier and angrier he grew, and it fell to Kamrushepa, goddess of healing and magic, to soothe him with her arts.

They finally induced Telepinu to return. And with him came a degree of normality. The Hittites saw that a staff had been thrust into the ground before the newly arrived Telepinu, and that a fleece was hanging from the staff. They took it as an omen: "It means the fatness of sheep, it means grains of wheat (and) grapes on the vine, it means cattle large (and) small, it means long years and offspring. It means the favourable message of the lamb… It means a fruitful breeze. It means… plenty."

We don't know how this myth concluded, but we can hazard a guess: doubtless with rain, because it always rains in the end. For the real-life Hittites, in spite of the mythical, drought-repelling bee, the rain possibly came too late. Scientists at Cornell University, analysing wood from a tomb thought to be that of King Midas's father, raised the question. They finally came to a conclusion that had already been hinted at by a previous study: a severe drought applied the final death blow when the Hittite Empire was already on its knees. To understand what clue the wood provided, we have to turn to a tree stump and a man named Andrew Ellicott Douglass.

In 1894, the world had just discovered the Hittite Empire, which had been completely forgotten since its sudden disappearance over 3,000 years before. Douglass, an astronomer and grandson of the man to first record a meteor shower in the US, was at work in Arizona on an assignment given to him by his boss, Percival Lowell. He was to find the best place to instal a telescope for observing Mars. Douglass did as instructed, found a good Mars-gazing spot, and construction began on the Lowell Observatory. From then on, things did not go smoothly with his boss. Lowell had set his heart on finding evidence of canals on Mars that would allow him to prove the existence of a civilization on the red planet, and in fact he wrote several books to back up his theory. He was not the only believer. A translation error in the work of an Italian astronomer turned *canali*—as in natural "channels" on the planet's surface—into "canals", spawning the idea, soon widespread in the US, of a Martian civilization not only blessed with intelligence but also thirsty, given that it had been forced to build canals for its limited remaining water. Something that afflicts many people on earth, projected onto other planets. This strange cosmic consolation also frightened people into expecting an alien invasion, although, if aliens did exist, they might simply have asked for a glass of water and gone home.

Douglass took issue with his boss's ideas and for that was fired after seven years at the company. He had lost his job, but not his curiosity, which he now trained on our own planet instead: he looked down at the ground and, his gaze meeting a tree stump, had an epiphany. There they were: the irregular rings, some thicker and others almost imperceptible, "spoke" to him. What did the tree want to tell him? Douglass still couldn't decipher its secret language, but the tree was telling him that there had been times of plenty when its growth had been unhindered, and other times when it had been an enormous effort just to survive.

The astronomer set out to understand the biography that the tree was showing him. He became obsessed with it because what he really wanted to show through tree rings was the influence of solar cycles on the Earth's climate. He knew that Leonardo da Vinci had been convinced that the thickness of these rings was dependent on atmospheric wetness. Douglass succeeded in dating trees based on their rings. Thus was born dendrochronology, which later gave rise to dendroclimatology. Douglass created the Laboratory of

Tree-Ring Research at the University of Arizona in 1937, but died without achieving his ultimate goal. Today, thanks to his epiphany or his moment of distraction, we have a better understanding of historical droughts and, to a certain extent, can predict future ones. This discipline, together with the study of Arctic polar ice caps, of pollen and marine sediments, of rogations recorded in church annals and of records of grape harvests, is the basis of almost everything we know about the droughts of recent centuries and one of our main tools for responding to droughts present and future.

Back to 2023, and Douglass's successors' research into the wood from the tomb with apparent links to King Midas. The uneven growth of the juniper tree from which it came provided clues to the fact that, for the Hittites, neither the bee nor the rain arrived in time. That no magic existed to bring back the rain when it was most needed. The abandonment of Hattusa was sudden and definitive. Those who did not die had to leave and journey on elsewhere.

Epilogue

Exodus of the thirsty

> These people feel no connection to the place in which they happen to be. They will soon depart, leaving no trace. In their ballads which they sing in the evenings, is a constantly repeated refrain: "My country: My country is where the rain falls."
>
> RYSZARD KAPUŚCIŃSKI,
> *THE SHADOW OF THE SUN*

> Slap-slap. His sandals flapped against the cracked earth.
>
> GRACILIANO RAMOS,
> *BARREN LIVES*

There is a thread that runs from us to the Neolithic farmers who scanned the sky for rain and to the Mesolithic hunter-gatherers who sheltered inside caves with views out over the water. This thread is made of esparto grass. Thousands of years ago, people searching for water discovered a grass that grew in dry lands and could provide them with fibres flexible enough to make ropes and baskets. Their descendants used it to make espadrilles which they placed in the Cave of the Bats (Granada) 6,200 years ago, probably as part of a funeral rite. Thanks to the dryness of both the surroundings and the inside of the cave itself, these have remained intact despite being made from once living matter. They are the oldest shoes ever found in Europe.

Over time, and despite the fact that the technique has hardly changed, esparto has been put to more and more complex uses and some of my ancestors, like those of a hundred years ago—country people whose existence differed little from that of Neolithic times—learned to make daily objects with it: saddlebags, chairs, panniers, mats, water bottles and cheese strainers. While my great-grandfather Pedro turned pumpkins into canteens, his sister Bernardina transformed esparto into baskets to rock her children between a couple of olive trees while she dug up onions in the Huerta Soriano, which she would later take to sell in neighbouring villages before going on to deliver letters to the prison in Villanueva de los Infantes. Her husband Nicolás was an inmate there; he was my great-grandfather but also my great-great-uncle because three offspring of my great-great-grandparents Norberto and Juana (Virginia, Nicolás and Gumersinda) had married three offspring of my great-great-grandparents José and Ángela (Pedro, Bernardina and Cruz).

Bernardina used esparto to weave a false bottom into her vegetable baskets to hide the letters. Every food delivery was also a way to communicate things that were for her husband's eyes only. Once her visit was over, he would chew through the cords to get to the letter. On one occasion, the basket she left behind contained a knife; unbeknownst to her, her daughter Juana had hidden it in the secret compartment. From then on Nicolás no longer had to use his teeth.

One day in May 1940, when she was no longer allowed to visit Nicolás, Bernardina received a letter from the prison in Ciudad Real, signed by him. He had encouraged day labourers in the village to demand their rights and to stand up to the local bosses, had founded a local trade union as well as Radio Comunista, and had also voluntarily enlisted in the Republican army. For all these reasons, and for contravening another Francoist rule by having music at the funeral of his mother—my great-great-grandmother Juana—Bernardina's husband had already been sentenced to death on the day she received the letter. So had her brother Cruz.

In that last letter, Nicolás offered Bernardina two pieces of advice: that she teach her children to write, and that she go wherever the need for water took her. "Bernardina," he wrote, "given what you've said about Amancio [their son] having no work because of the drought, I say that if things seem impossible to you there, go somewhere where you can live a better life." The first piece of advice, although meant only in relation to their own children, Bernardina took so seriously that she set up a school of sorts in her own house, teaching boys, girls, the elderly, mule drivers and landworkers how to read and write by the fireside. But the second part she ignored, however much her son also insisted they ought to move on.

So in the spring of 1940, barely a fortnight apart, brothers Nicolás and Cruz were lined up and shot. The pain of those losses would have been all the more, given the interconnectedness of the families. And as if that generation had not had it hard enough already—chipping away at the Minao rock in order to irrigate their crops, surviving the Spanish Civil War while having no idea where another of the siblings had wound up (he had gone to fight in Morocco), with various repressive measures that hit landworkers particularly hard and with having several deaths to mourn at once—drought then arrived.

There is a Spanish phrase that has come to be strongly associated with Franco, so often did he repeat it in his mellifluous tones. "Persistent drought" may just be his most famous saying, and it was his second great enemy after the "Judaeo-Masonic-Communist conspiracy". But he didn't coin it. A quick search in the newspaper archives of the National Library of Spain shows that the two words were already frequently used together in nineteenth-century print media, before Franco's bombastic cant commenced, before he was even born. Among the results thrown up by the search, the oldest mention of "persistent drought" dates from 1844 and appeared in a publication called *The Guide to Commerce*. Just five years later, it was already commonplace.

Five years after the end of the civil war, when the driest of those dry times began, my great-grandmother Virginia died for reasons that have never been entirely cleared up, but tend to be boiled down to "poverty". People point to malnutrition, and the illnesses that arise from it, an inability to pay for medication and some issue that resulted in respiratory problems. Her death certificate mentions renal failure. She was forty-two and had four children under eighteen when one of her brothers-in-law found her dead at home. Hundreds of thousands died in similar circumstances in the post-war years, with land workers in southern Spain in particular suspected of being unsympathetic to the regime.

According to Miguel Ángel del Arco, a researcher at the University of Granada, Franco hid the true cause of people's lack of food—his drive to make Spain self-sufficient—behind a façade of drought, war and international isolation. But here once more was a case of drought going hand-in-hand with despotism and dubious policies; the blame did not lie solely with the parched conditions, and yet these became a cover for an outright famine as well as a typhus epidemic caused by malnutrition, which, together with diseases such as smallpox, diphtheria, dysentery and typhoid, hit the rural south particularly hard.

Such ailments went undiagnosed in the most impoverished and isolated rural areas, where they often spread for want of the correct medicines, but among their main repercussions were kidney and lung failure, as is the case with other infectious diseases. Drought did not kill the woman for whom I am named. She was born into a family of poor labourers, the victims of reprisals, two of whose sons had been killed by firing squads in an area particularly harshly treated in the post-war period. As a result, she went hungry, fell ill with some unknown disease and had no access to treatment. Regardless of what her final symptoms were or the fact that she died the same year the Ebro and the Manzanares dried up, it seems clear that the real reason would not appear on a death certificate produced in this country in 1945. She died because she was poor and because her people were being roundly punished, not because it hadn't rained.

In his Christmas message in 1950, Franco said: "We have seen our land and our reservoirs dry up owing to persistent drought, which has had an unprecedented effect on industrial output." But Spain had not been afflicted by persistent drought from the exact moment he assumed power, nor were conditions worse than at other times in the same century when the outcome was not famine.

Although the records of the National Climate Data Bank of the State Meteorological Agency (AEMET) show that after 1940 the drought desisted and that it was not as prolonged as the one in the 1990s, that the driest years were between 1944 and 1946 and that, after showing a downward trend for more than half a century, in 2022 it rained on average twelve per cent less than in 1950, Franco clung to the idea of a "persistent drought" to justify his own decisions.

These data of course reflect the nation's overall annual rainfall, and there have often been droughts in one part of the Peninsula at the same time as floods in another. In 1940 Nicolás enjoined Bernardina to move somewhere less unforgiving, in 1945 the Ebro practically dried up and in 1947, when some places were beginning to recover, others were still experiencing major droughts. Data from the Ciudad Real observatory also indicate that 1954 was the year with the least rainfall in a province that normally does not even receive the La Mancha average of 400 litres per m² per year; the population of the island of El Hierro went through a period of great despair in 1948. Garoé, the sacred tree-fountain from which they believed rainwater to issue and which had supplied their Bimbache ancestors for centuries, failed them. That year, thousands of Canarians were forced to leave for Venezuela because of thirst.

At some point it dawned on Franco that persistent drought could provide him with what Manuel Vázquez Montalbán called his "meteorological alibi" to justify the famine as well as the country's glacial recovery after the civil war. It also enabled him to present

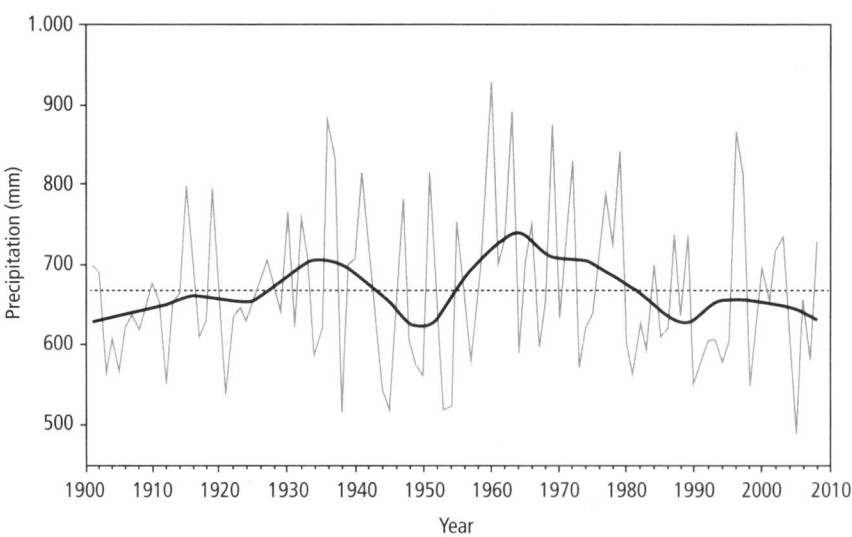

Overall annual precipitation in the twentieth century. Source: AEMET

himself as God's chosen one, destined to be the saviour of a country dying of thirst whose river water flowed irremediably to the sea. The empty reservoirs also provided the regime with material for its greatest PR campaign. It was so effective that you still hear some people parroting the idea that Spain has water to drink and irrigation for the crops thanks to a man who in the mid-twentieth century earned the nickname of Paco the Spider, along with jokes related to the endlessly repeated line: "This reservoir is now open." But the whole thing is an easily refutable fallacy.

Some of the reservoirs in Spain were built over 2,000 years ago, and there have been attempts to implement national water policies since the drought at the end of the 1800s. The reservoirs were convenient for a regime attempting to consolidate power: it could swiftly relaunch a fledgling industry that had been paralysed by the war, and also hide mass graves that no one would ever find again—unless drought, which now leads to submerged villages coming to light every single summer, became so intense that it ended up becoming an ally to historical memory. Those plans were not new, but were used to propagate false beliefs in a country that often forgets the blood and sweat that went into the building of the reservoirs on the part of people who were excluded from the inauguration photos and who swelled the ranks of displaced labourers. Among their number were immigrants from the most beleaguered areas—people who could only feed their families by taking construction work on the reservoirs, which would often go on to flood their own homes—and political prisoners.

In Spain there is such a long history of water inequality and related disagreements that our first recorded conflict was between two towns over the ownership of some water channels, as we have seen.

The Roman reservoirs in Proserpina and Cornalvo and the network of *qanats* in the Tabernas Desert, as well as the network of irrigation *acequias* in the Sierra Nevada and the Water Tribunal in Valencia, all bear witness to the ancient need to divert rivers and rain to areas of water scarcity. The Tibi reservoir has been in operation for 300 years. But it was particularly at the end of the nineteenth century that a national water policy, until then non-existent, became a pressing concern, as reflected in the newspapers of the day. People didn't give up on St Isidore, but the idea began to take hold that a living hero was needed to fight thirst.

As we have already seen, on 6th May 1896 nearly all of Spain was experiencing rain after a long drought. Despite scientific advances, some Spanish newspapers continued to credit Isidore, their rainmaker par excellence. The Little Ice Age had ended and a gradual warming was underway, but drought conditions never relented altogether. The saint was also asked to intercede in the war in Cuba—with Spain the victor, naturally.

So rain came and the Cuban war drew to an end. But there must have been some interference in the signals sent to the saint because his intervention didn't favour the people who had been doing the praying. From then on, and in order to overcome the collective grief at the loss of Spain's last overseas colonies, the source of all the country's ills had to be found. A culprit was soon identified: thirst. The water in the rivers was being wasted, people said, by being allowed to flow out to sea in a country where three-quarters of the rainfall was insufficient, late and erratic. This practical idea, already advocated by Don Quixote and by an earlier Spanish minister of finance, Juan Álvarez Mendizábal—that "doing good to villains is like throwing water into the sea" and "Spain will never be rich while its rivers flow into the sea"—became a national cause. The rivers had to be dammed and redirected. Politicians of Spain's *regeneracionismo* or regenerationism movement such as Joaquín Costa made the idea their rallying cry. Costa dreamed of a Ministry of Water, called for the nationalization of all water and proposed the colonization of Spain's interior as part of a plan to make better use of the rivers. He also believed that people should be moved down from the mountains into the valleys.

Regenerationist writings arose in parallel with a character named the "hydraulic apostle" by sociologist Alfonso Ortí, a man with the power to fix all water-related problems on a peninsula that, to make matters worse, was tilting. The incarnation of the hydraulic apostle would be another kind of rainmaker who would now use politics and engineering to carry out what James Scott called "state landscaping". This would involve fixing everything that, according to that perspective, nature (or God) had done wrong. If nobody had bothered to use a spirit level when it came to installing the Iberian landmass, the hydraulic apostle would see to it. His job was to plant trees to attract rain, build reservoirs to store it, instal irrigation and, at the same time, move people down from the mountains and settle them in empty, dry valleys so that they could work the land and make it just as productive as the rainy places. All this, which had already been proposed by the regenerationists, was implemented by Franco's regime half a century later.

The nineteenth century in Spain ended with rogations, with various legislative proposals to make better use of water resources and with the creation of a national water company. The hydraulic apostle that Spain so longed for was first embodied by Costa and culminated in Franco, but between the policy proposals of the regenerationists and the National Public Works Plan of 1940, there were attempts such as the National Plan for Irrigation Channels and Reservoirs of

1902 (better known as the Gasset Plan), the Irrigation Act of 1911 and the National Plan for Hydraulic Works of 1933.

A subsequent drought, between 1912 and 1914, turned the river I can see from where I write into a sorry trickle and drove much of the rural population of Aragón to seek work elsewhere. An exodus also took place of people from Zaragoza and Barcelona to countries like Argentina and Panama. As if that were not enough, grape phylloxera had already been destroying vines in Spain and parts of France for forty years, driving farmers from their land. Phylloxera would probably not have spread to such an extent in Spain without the concurrent drought; there was a view among many farmers who had invested in vines just as phylloxera was decimating them in France that the problem was simply the lack of rain. A belief they held for as long as it took the parasite to destroy a vineyard and gain unstoppable momentum, which is the same time as a drought lasts. Once the carrier insect had been identified, it was already too late.

In the Valencian village of Aielo de Malferit there is a story about one of its residents, Bautista Aparici, who took desperate measures and, on travels in the US, began buying healthy vines in exchange for a tonic he had invented together with two other young men from the village. A gesture like this speaks to the hopelessness that must have taken hold when practically all the vineyards had been wiped out, as it was precisely from the US that the aphid in question had arrived, spreading from France to Spain and the rest of Europe. But that was the solution they resorted to in other severely affected areas too, such as Valdepeñas, where there is evidence of grapevine cultivation dating back thousands of years.

The beverage in question soon gained in popularity thanks to Aparici's market-expanding travels, while Ricardo Sanz and Enrique

Ortiz stayed behind manning the bottling plant. It was called Nuez de Kola-coca, it was made from kola nuts, Peruvian coca and water, and it was both thirst-quenching and soothing for stomach aches. There are labels for Nuez de Kola-coca dating back to at least 1882. The drink was even entered into a competition in Philadelphia in 1885, where it won a medal, as it had done at other trade fairs around the world. A year later, a pharmacist from Georgia by the name of John Stith Pemberton patented Coca-Cola, in a repeat of another invention taken from something the Mayans had also used to quench their thirst: chewing gum. What could explain such a coincidence? Coca leaves had been used since time immemorial to relieve stomach pain. Pemberton had been looking for a way to combat the morphine addiction he had been left with following medical treatment for a sabre wound sustained in the American Civil War. Once the war was over, he invested most of his money in the search for the perfect elixir, and experimented with various ingredients. His drink, made with wine, coca and damiana, became very popular, but failed to ease his pain. To make matters worse, alcohol was banned in his home city of Atlanta in 1886. With the help of a friend, he replaced the wine with sugar syrup, the damiana with kola nuts and, by mistake, he also added some carbonated water. The result became a popular non-alcoholic substitute for the tonic he had previously begun selling; workers drank it to help them through the exhausting days. One of Pemberton's partners came up with the name Coca-Cola. But he never got to enjoy the success of his invention: two years after coming up with the formula, his morphine addiction led him to sell the rights, and shortly afterwards he died of stomach cancer, destitute and never getting to see the popularity of the drink, which had more to do with thirst than with stomachache.

During those years, with Europe's grapes withering on the vine, Spanish farmers who did not migrate to America out of thirst did so because of phylloxera, the problem of which was aggravated in places by the quasi-feudal system of *caciquismo*. At that time, three-quarters of Spaniards lived in rural places, and the League for the Relief of the Destitute, a forerunner of today's food banks, was set up to support migrant pickers. Those who did not go to seek a new life in the factories of nascent industries in Catalonia or the Basque Country boarded a ship. Between 1910 and 1929, the Spanish population arriving in Argentina tripled. From the Marina Alta in Alicante alone, 10,000 rural workers left for the US and Canada.

Reservoirs were also planned and built at that time. When the *Lusitania* sank in 1915 and Frederick Starck Pearson—the Canadian founder of the Barcelona Traction, Light and Power Company which electrified Catalonia's capital—was among those who drowned, he had already planned several dams along the Ebro River in Aragón. King Alfonso XIII made appearances at reservoir construction sites in the 1920s, such as the Ebro (Cantabria) and El Chorro (Málaga). Shortly afterwards, he smoked a Havana cigar upon inaugurating the Guadalmellato reservoir (Córdoba), on a Moorish religious site that has now been uncovered as a result of drought. The King's Little Path (Caminito del Rey), known today as one of the most vertiginous mountain passes, is so called because Alfonso walked along it to inaugurate the Guadalhorce-Guadalteba reservoirs (Málaga). Although in reality, that suspension footbridge had been built at the turn of the century to link two reservoirs.

In the 1930s, Spain was once again struck by drought, particularly in Andalusia and La Mancha. Such was the impact on Terrinches that not even the descendants of those who lived through it have forgotten: women formed long queues to pass basins of water from a spring that had dried to a trickle. According to the old people in the village today, those who were elderly then spoke of it as the worst drought in living memory. Rural people of La Mancha and Andalusia had to sow in cracked fields and slaughter lambs at birth, impossible as it would have been to raise the animals in such conditions, and many of them were doubtless also forced to emigrate. "For eight days of rain I would give the 100 million pesetas I am being offered by the Minister of Finance, but the only showers you get here are the politicians," said the Minister of Finance Luis Rodríguez de Viguri after lamenting that particular "persistent drought", as reported in *El Mundo* on the 24th of November 1930. Articles in the journal *La Industria Pecuaria* discussed the lack of rain for months, from summer to December. The year ended and still there was no rain.

Indalecio Prieto, Minister of Public Works during the Second Republic, created the Centre for Hydrographic Studies, attached to the Directorate General of Hydraulic Works. He commissioned its chief engineer, Lorenzo Pardo, to draw up a national water works plan. Pardo, who had been a key player since his role in launching the Ebro Hydrographic Union Confederation (later the Ebro Hydrographic Confederation), believed that water policy up to then had been blighted by inequality and limited scope. The prevailing ideas up to that time were essentially as follows: irrigation is always a good thing, just because it is, in all places, for everyone and at any cost. The particularities of respective catchments were not taken into account until the creation of the hydrographic confederations, which were initially independent of the state and included the participation of farmers. Pardo's theory of "hydraulic disequilibrium" held that three quarters of the country comprised arid or semi-arid terrain. He concluded that, although the rivers on the Atlantic-facing parts carried more water, the land there was less productive than the Mediterranean slopes, where the relative lack of river water was mitigated by superior irrigation. His studies led him to view the Mediterranean zone as better suited for export crops, and the Atlantic zone for domestic consumption.

With his national plan, he believed he had improved on policies that had treated the whole country as homogenous. But it involved diverting rivers such as the Tagus and the Guadiana to boost irrigation in places like Castellón and Almería. Although Pardo had initially won the admiration of farmers by inviting their input, the Júcar farmers were soon up in arms over these proposed diversions. Moreover, he was accused of not considering the social repercussions, of favouring the Mediterranean and of forgetting all about Castilla. The latter was, in fact, a sentiment deeply rooted in the central Meseta (the Inner Plateau) ever since the regenerationist attempts to save Spain by damming rivers. Many believed the project was "neither a plan nor national" and that it would only benefit the wealthy, at the expense of those already blessed with fewer resources.

While it is certainly true that most of the reservoirs in Spain—nowadays including Europe's largest—were built under Franco, many had been slated or built, or were under construction, before then. Hence in 1933, the same year Pardo drew up his long-term plan, the national press could already boast about the largest reservoir in Europe, the Jándula dam in Jaén.

The construction of the Fuensanta reservoir in Albacete and the inequitable distribution of nearby land in 1933, along with the resultant penury for so many in the surrounding countryside (once again, the endemic feudalism was also to blame), led to clashes between farmers and the Guardia Civil in Yeste, in which eighteen civilians and one member of the Guardia Civil died. This was in the days before the coup d'état that sparked the civil war, and some historians consider this event to have lit the fuse. The National Water Works Plan, which had been making baby steps, was then brought to an abrupt halt by the war. In 1940, not a year after the end of the war, the Public Works Plan was approved; on matters of water it shared many similarities with Pardo's aborted project. When Franco inaugurated the Ebro reservoir in Cantabria in the summer of 1952, it was thirty-one years after the first stone had been laid. Pardo himself had put forward blueprints for Ebro in 1916. The works were completed decades later thanks to the slave labour of hundreds of political prisoners.

Candido stood looking at the feet of the workers on the coffee plantation where he had been born. They looked like misshapen maps to him, and he felt like they had a story to tell. In the *sertão* ("hinterland" or "backcountry") of northeastern Brazil where this nine-year-old boy had grown up painting, he had become accustomed to the way his people traipsed along every time another drought forced them to relocate. Drought was a cyclical fact of life. A long time after that, he began making paintings of those thirsty Brazilian exiles who left such an impression on him during the drought of 1915. The workers at the centre of pictures like *The Migrants* and *Dead Child* were intentionally colossal in size. And they had misshapen feet that told a story.

It took Candido Portinari time to begin exhibiting because when he started out as an artist, the Brazilian government was intent on propagating an image of the country as a verdant orchard, and it censored all depictions of drought. Until suddenly there was an unstoppable artistic explosion, like a river bursting through a dam, that produced paintings, songs and books—some of which were also made into films—and turned thirsty migrants into the true protagonists of the twentieth century. Portinari's migrants were like tumbleweed in the way they moved about so much—and they weren't the only ones.

The Great Plains of the central US saw no rain in 1930. In the midst of the Great Depression, the worst drought in a thousand years was accompanied by dust storms, tornadoes, snow and creditors wanting their money back. Over the course of a decade, the "black blizzards" or "black rollers" claimed millions of lives, as hectare upon hectare of corn and wheat were turned to desert. There were days when visibility was reduced to a metre ahead and the noise of the roaring wind was compared to "a truck going up a mountain in second gear".

After the dust had swallowed their homes and crops and the banks had taken their property, they were sold a story of a promised land to the west in California, and hundreds of thousands left Oklahoma, driven out by thirst, just as had happened thousands of years before in Caral and Akkad. That episode, as well as the geographical region it encompassed (Oklahoma, Kansas, Texas, Nebraska, South Dakota and Colorado), became known as the Dust Bowl, and the thirst-displaced people lured to California under false pretences were called Okies, Oklahoma being their epicentre (although the term has since become derogatory).

In 1936, Franklin D. Roosevelt created the Soil Erosion Service and commissioned a report to determine the causes. The committee concluded that the Dust Bowl was largely man-made. Chief among its causes were environmental policies that pushed unskilled farmers to overexploit the land, decimating it with over half a century of unremitting ploughing. The Homestead Act, which in 1862 provided arable land to settlers who moved en masse, was deemed to have been a major failure with catastrophic consequences.

In the summer of 1936, at the same moment as the military uprising that preceded the Spanish Civil War, John Steinbeck spent time with some of the Okies on the roadsides where they set up their shacks, and his reports on their tribulations were published in *The San Francisco News*. Those interviewees were the inspiration for the Joads, the fictional family that was the subject of his most famous novel.

In one of the most well-known lines from *The Grapes of Wrath*, Steinbeck describes man on the basis of hunger. But it was thirst that originally combined with the Great Depression and creditors to drive those people into vagabondage.

According to a UNESCO report, in the last decade alone more than 260 million people have either had to migrate, been displaced or lost their homes due to climate disasters, mostly as a result of global warming. It is estimated that the number of climate refugees will continue to rise and that drought will be one of the principal factors. "Of the estimated one billion-plus migrants in the world, at least ten per cent are motivated to seek a better life elsewhere because of water scarcity", according to a World Bank report presented at World Water Week in Stockholm.

"Cli-migration" is a new word, although the concept is not: peoples have forever been displaced because of the climate. Cli-migrants have included those of our ancestors who fled Africa; they can be considered responsible for bringing ancient Egypt and Sumer into being; and it is of course because of them that Spanish has so many words rooted in Arabic. According to recent research, sixty per cent of Spanish twenty-somethings already assume that they will have to move abroad to escape global warming, which, if the forecasts are correct, is due to scorch these lands to the point of virtual uninhabitability. The paradox, as we have seen, is that in extreme climatic situations the Iberian peninsula was once the last refuge on the continent. But this happened before we altered what would be the Earth's natural trends, if only non-human causes were at work.

We need new words, because what is not named does not exist. What to call people forced into exile by thirst? They (or we) are going to become more and more numerous over time, and we have to start naming them, as was done with the Okies. I would suggest "hydro-migrants", because that would also encompass the tens of thousands displaced by large hydraulic projects around the world, either when the local population is evicted or when labour is brought in from poorer areas, which are often also very dry areas.

They are also united by a feeling. "They are resourceful and intelligent Americans who have gone through the hell of the drought, have seen their lands wither and die and the topsoil blow away; and this, to a man who has owned his land, is a curious and terrible pain," Steinbeck said of the Okies. There was no word at the time for that pain, but one has now been coined by the philosopher Glenn Albrecht: "solastalgia". This is the distress caused by environmental change. That is to say, those who flee because of thirst in the Horn of Africa or the Great Sertão share the pain of people in León or Aragón who have seen their homes engulfed by reservoirs. Solastalgia has a future equivalent that becomes a specific type of fear: "eco-anxiety".

To travel through the ruins of Spain's submerged places is to hear stories of old people who died of grief, and of depression and anxiety when these things had no names, or because people killed themselves when suicide too was something unspeakable. In Argusino (Zamora) I went to see the remnants of a cemetery that had been flooded for a reservoir and later become accessible again when it dried up; a local man talked to me of his father's depression, pointing out the ruins of his house, and of neighbours who had taken their own lives. A couple of days later I met a man who had burned his own house down in Riaño before it could be flooded. He still didn't want to—or couldn't—go back there in his mind. There was a famous case of a man called Simón Pardo who shot himself on being told he was about to lose his home; that became a media story, but there were also those who died of heart attacks on hearing of the plans. Weeks after that trip, I told these stories at a book launch in Mequinenza (Zaragoza). One elderly lady, who made several comments without taking off her sunglasses, finally said: "It really was like that. I lived through it all here, just like them, from beginning to end. And my father was one of those who tried to commit suicide. One day, they found him on the bridge saying he was going to throw himself into the river." For a long time, we have seen all that sacrifice—which has meant we've been able to hydrate, to take showers, water our cracked gardens, to turn on the lights and charge our mobile phones—and looked the other way. We have a long-standing debt to those who really were the ones to allow us to quench our thirst.

Thirst was also responsible for sending the boy I spoke about in the prologue up on to a roof in Riaño in the time after Spain's transition to democracy. I didn't know about this town, I said, because I had only just been born then, and because the suffering of so many was quickly forgotten. But I didn't want to finish without learning more about it. In recent years I made several trips to that and other areas where villages were flooded and inhabitants evicted, and

I interviewed some of the people who were forced to leave their homes to make way for reservoirs or eucalyptus plantations designed to prevent silting, people who were both expelled from a dying land and went on to provide the labour in the building of the reservoirs. One of them was José Francisco, to whom my ignorance in a sense indebted me, because those who search for water and those whose homes have been submerged by it are, deep down, one and the same.

José Francisco was waiting for me in a chapel full of former Riaño inhabitants; they had rebuilt the chapel at a higher point, moving it one stone at a time to save it from being flooded. I had a hard time recognizing him: it had been thirty-five years since the Mauricio Peña photograph. Although after the flooding a new village had been built, also called Riaño but nowadays part of the so-called "fjords of León", most of the former inhabitants had dispersed across the Peninsula and even farther afield. José Francisco now lives in Valladolid, but he visits the chapel every year, along with those other former locals. The famous photograph featured more than just the boy in the cowboy hat with his pitchfork. There is also a man lying on the roof. This is his uncle.

After days of holding out, along with other locals and activists who had come to support them, seeing that no more could be done, they got down from the roof. It was just before this that Mauricio Peña took his photograph. As if in a scene from Valentin Rasputin's classic *Farewell to Matyora*, which is concerned with the devastating impact of industrialization and urbanization on peasant life, the uncle went back to his house and set fire to it. The nephew went up to the attic to pick out some final object to take with him before the waters came. Choosing a toy fort without thinking too much, he clasped it to himself and left the house for the last time. At the chapel he told me that, once installed in his new house, he put it up in the attic there as well, but unfortunately there was a leak. Every time it rained, a drop—never the same drop, though it seemed like it was—would land on the toy fort, until eventually it disintegrated. The lives of the drowned and the thirsty are full of such paradoxes. There is something fate-like about it: water, or its absence, pursuing you everywhere you go. Arundhati Roy has written about those who enter a loop from which they find it difficult to escape. "Some of them have subsequently been displaced three and four times—a dam, an artillery proof range, another dam, a uranium mine, a power project. Once they start rolling, there's no resting place. The great majority is eventually absorbed into slums on the periphery of our great cities, where it coalesces into an immense pool of cheap construction labour (that builds more projects that displace more people)."

The International Commission on Large Dams estimated that between forty and eighty million people have been displaced by this type of construction worldwide. But the figures don't add up: forty million is the approximate number of people displaced in India alone, while in China the Three Gorges Dam displaced 1.5 million people, and Egypt's Aswan Dam displaced between 60,000 and 90,000 people. And there are already 45,000 large dams worldwide, not counting smaller levees and river barriers. To say nothing of the incalculable number of people forced to leave when their crops or the villages on which they depended economically are flooded, or when trees are planted to prevent reservoirs silting up, to encourage rainfall or to help the soil absorb water. In Spain, the people exiled because of thirst are not only the 50,000 or so counted as having been forced to evacuate their houses, they are also all those displaced because of reforestation, irrigation settlers (who are often the same people) and construction workers on reservoirs who risked life and limb, as well as the residents of nearby villages whose livelihoods were taken away. This is why they can never be quantified. How can we put a number on those who were sold the panacea of a new town, house and land, land that was hard to grow on and that took them so many years to pay for? Those who were "given the opportunity" to buy back their houses—now ruins in places with no services—after they were expropriated? Those who were displaced by reservoirs that never got built or who lost their livelihoods to the water but not their villages?

That uprooting was the price people have paid for thirst. Promoted to enable irrigation in places where it did not rain, to repopulate empty lands, to feed industry and bring running water and electricity to homes, reservoirs soon became, together with trees, modern talismans for invoking rain. They were sacred and their promoters were the embodiments of the new hydraulic apostles. Such is the fear of dying of thirst.

Several hunger stones recently surfaced in central Europe. During the Little Ice Age, it became popular there to engrave river rocks with the dates of great droughts, accompanied with warnings: "If you see me, weep," reads the most famous. At times the water covers these hydrological markers, known as Hungerstein, and at others it leaves them exposed, allowing the dead to send messages to the living. But sometimes it is the other way around, and the living communicate with the dead through these time capsules written centuries ago. Recently, someone even dared to engrave a reply: "Don't worry, girl, and don't cry. Just water your field when it dries out." This more recent engraver knew that the stone's re-emergence was not such a catastrophe given that, since the construction of a dam in the early twentieth century, it has been visible for some 126 days every year. There is even a legend surrounding that stone which speaks to the sacredness of reservoirs; if another dam that has been planned since 1653 is finally built in Decin, it says, the stone will never be seen again and there will be no more droughts. As it claims on another of the Hungerstein: "Life will blossom again once this stone disappears."

It's raining. Since I started this book, I have been hoping that rain might come on the day I reach the final page. I wasn't overly hopeful, in truth; we are in the second consecutive drought year in a country that suffers three-year droughts practically once a decade. There was hardly a drop this spring. But I was woken by rain this morning and now, several hours later, it is still falling as I write, door flung open to let the petrichor waft in. This sensation as I smell the wet earth joins me to Lucy and to Eve, and to Manuela as well, a woman born midway through the seventeenth century in Terrinches and the earliest of my ancestors I've been able to uncover in the church records. When it rains there, it doesn't only mean that it's a good day for making pease pudding, but that we can soon go and see how the Cañico spring has responded. This is the true measure of the village's wellbeing, and possibly what those of us who are far away miss the most.

ACKNOWLEDGEMENTS

Certain people were particularly significant for me during the drought that was the seed for this book, and during its writing as well. My parents, my brother, my grandparents, Paula, David and the two Antonios, father and son. I don't know if I would have written again if it hadn't been for the person who accompanies me, supports me and pushes me to places I don't even think I am capable of reaching. Thanks to Dani for the constant support and for putting up with the "thirsty" discoveries I tell him about every day as if I were a child discovering the world for the first time.

Infinite thanks to Ella Sher, my agent, who was such a believer in this project. Without her contagious enthusiasm, a book that was barely more than the beginnings of an idea would never have been picked up by several Spanish and Italian publishers. I thank those publishers for the impetus they gave me. My special thanks to Elena Martínez, Nacho Ruiz and Paloma Abad, my publishers in Spain, and to the rest of the publishers and translators making it possible for this book to reach such unexpected places as Germany, the UK, Italy, the Netherlands, Portugal, Mexico, Colombia and Panama: Simon Lörsch, Maria Meinel, Adam Freudenheim, Thomas Bunstead, Vicki Berwick, Rory Williamson, Andrea Tramontana, Elisa Tramontin, Jacoba Casier, Arieke Kroes, Eurídice Gomes,

Isabel Fausto... Thanks also to Irene and all at Penguin who have been involved, from proofreaders to the people in marketing. Thanks also to Pilar Álvarez for pushing me to tell this story. And to Julio Llamazares, even though I didn't do exactly as he suggested because I got sidetracked by thirst.

My thanks to family and neighbours in Terrinches and Villanueva de la Fuente who answered my questions and shared their memories and information with me. In addition to my parents, my grandmother and my brother, I am also indebted to my Aunt Paula, my Aunt Paqui, my Great-Aunt Angela, Francisco Javier, Nicasio, María Cruz, Juan de Dios, Pili, Juanjo, Nicolás, María Dolores, Juanvi, Paula, David, Ángel, Inma and Joaquín. Thanks to those who helped me with the details about the water war in Villanueva de la Fuente, such as Juan Ángel and Daniel. To those who are no longer with us but who told me, years ago, some of the stories that I have included here, such as Juan the miller and my Uncle Amancio. Thanks to those who have helped me, with their respective professional expertise, to clarify some details, such as Natalia (beekeeper), María Ángeles (doctor) and María (cheesemaker).

Special thanks to Luis Benítez de Lugo, who has dedicated a large part of his life to proving that we people of La Mancha did have a prehistory and that it was also characterized by thirst. He went with me to Castillejo del Bonete and to the Motilla del Acequión, he has spent years answering all my questions about Motillas and Yamnaya and he was one of the early readers of this book. Thanks also to Miguel Torres, who went with me to the Motilla del Azuer, and to other archaeologists, historians and anthropologists who have answered my questions or given me leads, such as Paul Preston, Honorio Javier Álvarez, Esther Rodríguez, Fernando Domínguez, María G. Alonso, Ramón J. Soria and my former anthropology

ACKNOWLEDGEMENTS

professor, Jordi Ferrús. Thanks to the scholars who, without falling into pure environmental determinism, have none the less avoided denying the impact of the climate. Reading their work has not only enriched me greatly, but also helped me realize that when I set out the main thesis of this book I had not gone completely mad. Their names appear in the bibliography.

Without the help of physicist José Javier Ruiz and the astronomer Jorge Gómez, I don't know if I would have been able to easily describe the formation of the Earth, the variations in its position over time and why long-term climate change is the price we pay for getting to live here.

Thanks also to my current neighbours in Castelserás, whether they really threw the Christ figure into the river or whether it was a sack of straw. I am especially grateful to Pilar, María José and Esther for not throwing me in too, and for sharing the story, real or made-up, of their most famous rogation.

Thanks to Doña Vicenta for teaching me how to read and write. Thanks to Fay for taking care of my eyes. Thanks to Elena and Merche, librarians from Alcañiz, for providing me with books and for keeping me company when I'm not writing. Thanks to Eugenio, Gala and Paula, the booksellers who kindly ordered all the books I asked them for and whom I now count as colleagues. And thanks to Antonio at the Spanish National Library, who always ends up giving me a hand with the newspaper and periodicals library. And of course, thanks to those who read excerpts before the book existed: Dani, Laura, Esteban, Jorge, Luis and Richard. Thanks to Gema, from Nuberia, who made me a beautiful pendant in the shape of a divining rod so that I could carry with me the questions that my grandparents are no longer around to answer.

SELECT BIBLIOGRAPHY

Mary Austin, *The Land of Little Rain*, New York: Houghton, 1903.

Miguel de Cervantes, tr. John Ormsby, *The History of Don Quixote*, New York: Norton, 1981.

Lewis Dartnell, *Origins*, London: Bodley Head, 2019.

Felipe Fernández-Armesto, *Civilizations*, London: Macmillan, 2000.

James Frazer, *The Golden Bough*, London: Penguin, 2006

Elena Garro, tr. Ruth L.C. Simms, *Recollections of Things to Come*, Austin: University of Texas Press, 1969.

Núria Bendicho Giró, tr. Maruxa Relaño and Martha Tennent, *Dead Lands*, London: 3TimesRebel Press, 2022.

Marcel Griaule, *Conversations with Ogotemmêli: an introduction to Dogon religious ideas*, London: International African Institute, Oxford University Press, 1965.

Richard Hamblyn, *The Invention of Clouds*, London: Picador, 2002.

Herodotus, tr. A.D. Godley, *Histories*, Cambridge, Mass.: Harvard University Press, 1920.

Luke Howard, *On the modification of clouds and on the Principles of their Production, Suspension, and Destruction: Being the Substance of an Essay read before the Askensian Society in the Session 1802–3*, London: Talor, 1830.

Donald C. Johanson and Edey Maitland, *Lucy, the Beginnings of Humankind*, New York: Simon and Schuster, 1990.

Ryszard Kapuściński, tr. Klara Glowczewska, *The Shadow of the Sun*, New York: Alfred Knopf, 2001.

Graciliano Ramos, *Barren Lives*, Austin: University of Texas Press, 1965.

Arundhati Roy, "On the Narmada Resistance", *Outlook Magazine*: April 1999.

John Steinbeck, *The Grapes of Wrath*, London: Penguin Classics, 2000.

___, *The Harvest Gypsies: On the road to the Grapes of Wrath*, Berkeley: Heyday Books, 1988.

Karl Wittfogel, *Oriental Despotisms*, New Haven: Yale University Press, 1957.

AVAILABLE AND COMING SOON FROM PUSHKIN PRESS

Pushkin Press was founded in 1997, and publishes novels, essays, memoirs, children's books—everything from timeless classics to the urgent and contemporary.

Our books represent exciting, high-quality writing from around the world: we publish some of the twentieth century's most widely acclaimed, brilliant authors such as Stefan Zweig, Yasushi Inoue, Teffi, Antal Szerb, Gerard Reve and Elsa Morante, as well as compelling and award-winning contemporary writers, including Dorthe Nors, Edith Pearlman, Perumal Murugan, Ayelet Gundar-Goshen and Chigozie Obioma.

Pushkin Press publishes the world's best stories, to be read and read again. To discover more, visit www.pushkinpress.com.

I LIVE A LIFE LIKE YOURS
JAN GRUE

A LINE IN THE WORLD
DORTHE NORS

STALKING THE ATOMIC CITY
MARKIYAN KAMYSH

CLOUDS OVER PARIS
FELIX HARTLAUB

THE WOLF AGE
TORE SKEIE

A WOMAN IN THE POLAR NIGHT
CHRISTIANE RITTER

A LIFE IN THE MAKING
FRANZ MICHAEL FELDER

MAZEL TOV
J.S. MARGOT

DAYS IN THE CAUCASUS
BANINE

ON LOVE AND TYRANNY
ANN HEBERLEIN

THOSE WHO FORGET
GÉRALDINE SCHWARZ

YOUNG REMBRANDT
ONNO BLOM

THE WORLD OF YESTERDAY
STEFAN ZWEIG

NO PLACE TO LAY ONE'S HEAD
FRANÇOISE FRENKEL

DREAMERS
VOLKER WEIDERMANN

THE LIMITS OF MY LANGUAGE
EVA MEIJER

A CHILL IN THE AIR
IRIS ORIGO

RED LOVE
MAXIM LEO

A WORLD GONE MAD
ASTRID LINDGREN

ON THE END OF THE WORLD
JOSEPH ROTH

SORROW OF THE EARTH
ERIC VUILLARD

A SORROW BEYOND DREAMS
PETER HANDKE

MEMORIES: FROM MOSCOW TO THE BLACK SEA
TEFFI